1977

THE
PHENOMENON
OF SCIENCE

THE PHENOMENON OF SCIENCE

V. F. TURCHIN

Translated by Brand Frentz

New York ■ Columbia University Press ■ 1977

Library of Congress Cataloging in Publication Data

Turchin, Valentin Fedorovich.

The phenomenon of science.

Includes bibliographical references and index.

1. Science—Philosophy. 2. Evolution. 3. Cosmol-
ogy. 4. Cybernetics. I. Title.

Q175.T7913   501   77-4330

ISBN 0-231-03983-2

New York ■ Columbia University Press ■ Guildford, Surrey

Copyright © 1977 by Columbia University Press

All Rights Reserved

Printed in the United States of America

Contents

Foreword

VALENTIN TURCHIN presents in *The Phenomenon of Science* an evolutionary scheme of the universe—one that begins on the level of individual atoms and molecules, continues through the origin of life and the development of plants and animals, reaches the level of man and self-consciousness, and develops further in the intellectual creations of man, particularly in scientific knowledge. He does not see this development as a purposeful or preordained one, since he accepts entirely the Darwinian law of trial and error. Selection occurs within a set of random variations, and survival of forms is a happenstance of the relationship between particular forms and particular environments. Thus, there are no goals in evolution. Nonetheless, there are discernible patterns and, indeed, there is a "law of evolution" by which one can explain the emergence of forms capable of activities which are truly novel. This law is one of the formation of higher and higher levels of cybernetic control. The nodal points of evolution for Turchin are the moments when the most recent and highest controlling subsystem of a large system is integrated into a metasystem and brought under a yet higher form of control. Examples of such transitions are the origin of life, the emergence of individual self-consciousness, the appearance of language, and the development of the scientific method.

Many authors in the last century have attempted to sketch schemes of cosmic evolution, and Turchin's version will evoke memories in the minds of his readers. The names of Spencer, Haeckel, Huxley, Engels, Morgan, Bergson, Teilhard de Chardin, Vernadsky,

Bogdanov, Oparin, Wiener and many others serve as labels for concepts similar to some of those discussed by Turchin. Furthermore, it is clear that Turchin knows many of these authors, borrows from some of them, and cites them for their achievements. It is probably not an accident that the title of Turchin's book, "The Phenomenon of Science," closely parallels the title of Teilhard's, "The Phenomenon of Man." Yet it is equally clear that Turchin does not agree entirely with any of these authors, and his debts to them are fragmentary and selective. Many of them assigned a place either to vitalistic or to theological elements in their evolutionary schemes, both of which Turchin rejects. Others relied heavily on mechanistic, reductionist principles which left no room for the qualitatively new levels of biological and social orders that are so important to Turchin. And all of them—with the possible exception of Wiener, who left no comprehensive analysis of evolution—wrote at a time when it was impossible to incorporate information theory into their accounts.

The two aspects of Turchin's scheme of cosmic evolution which distinguish it from its well-known predecessors are its heavy reliance on cybernetics and its inclusion of the development of scientific thought in evolutionary development that begins with the inorganic world. The first aspect is one which is intimately tied to Turchin's own field of specialization, since for many years he was a leader in the theory and design of Soviet computer systems and is the author of a system of computer language. Turchin believes that he gained insights from this experience that lead to a much more rigorous discussion of evolution than those of his predecessors. The second aspect of Turchin's account—the treatment of scientific concepts as "objects" governed by the same evolutionary regularities as chemical and biological entities—is likely to raise objections among some readers. Although this approach is also not entirely original—one thinks of some of the writings of Stephen Toulmin, for example—I know of no other author who has attempted to integrate science so thoroughly into a scheme of the evolution of physical and biological nature. Taking a thoroughly cybernetic view, Turchin maintains that it is not the "substance" of the entities being described that matters, but their principles of organization.

For the person seeking to analyze the essential characteristics of Turchin's system of explanation, two of his terms will attract attention: "representation" and "metasystem transition." Without a clear understanding of what he means by these terms, one cannot comprehend the overall developmental picture he presents. A central issue for critics will be whether a clear understanding of these terms can be gained from the material presented here.

One of the most difficult tasks for Mr. Frentz, the translator, was connected with one of these central terms. This problem of finding an English word for the Russian term *predstavlenie* was eventually resolved by using the term "representation." In my opinion, the difficulty for the translator was not simply a linguistic one, but involved a fundamental, unresolved philosophical issue. The term *predstavlenie* is used by Turchin to mean "an image or a representation of a part of reality." It plays a crucial role in describing the situations in which an organism compares a given circumstance with one that is optimal from the standpoint of its survival. Thus, Turchin, after introducing this term, speaks of a hypothetical animal that "loves a temperature of 16 degrees Centigrade" and has a representation of this wonderful situation in the form of the frequency of impulses of neurons. The animal, therefore, attempts to bring the given circumstances closer and closer into correspondence with its neuronal representation by moving about in water of different temperatures.

This same term *predstavlenie* is also used to describe human behavior where the term "mental image" would seem to be a more felicitous translation. If we look in a good Russian–English dictionary, we shall find *predstavlenie* defined as "presentation, idea, notion, representation." At first Dr. Turchin, who knows English well and was consulted by the translator, preferred the translation "notion." Yet it seemed rather odd, even vaguely anthropomorphic, to attribute a "notion" to a primitive organism, an amoeba, or even a fish. On the other hand, the term "representation" seemed too rudimentary for human behavior where "idea" or "mental image" was clearly preferable.

This difficulty arose from the effort to carry a constant term through evolutionary stages in which Turchin sees the emergence of qualitatively new properties. The problem is, therefore, only second-arily one of language. The basic issue is the familiar one of reduc-tionism and nonreductionism in descriptions of biological and psy-chological phenomena. Since the Russian language happens to possess a term that fits these different stages better than English, we might do better to retain the Russian *predstavlenie*. In this text for a wide circle of English readers, however, the translator chose the word "representation," probably the best that can be done.

The difficulties of understanding the term "metasystem transi-tion" arise from its inclusion of a particular interpretation of logical attributes and relations. Turchin believes that it is impossible to de-scribe the process by which a particular system develops into a metasystem in the terms of classical logic. Classical logic, he says, describes only attributes, not relations. For an adequate description of relations, one must rely on the Hegelian dialectic, which permits one to see that the whole of a metasystem is greater than the sum of its subsystems. The Hegelian concept of quantitative change leading to qualitative change is thus not only explicitly contained within Tur-chin's scheme, but plays an essential role in it. The behavior of human society is qualitatively different from the behavior of individ-ual humans. And social integration, through the "law of branching growth of the penultimate level," may lead eventually to a concept of "The Super-Being."

These concepts show some affinities to Marxist dialectical mate-rialism, in which a similar differentiation of qualitatively distinct evolutionary levels has long been a characteristic feature. The British scientist J. D. Bernal once went so far as to claim that this concept of dialectical levels of natural laws was uniquely Marxist, when he wrote about "the truth of different laws for different levels, an essen-tially Marxist idea." However, many non-Marxists have also ad-vanced such a view of irreducible levels of laws; one should there-fore be careful about terming a system of thought Marxist simply because it possesses this feature. Most Marxists would reject, at a minimum, Turchin's discussion of the concept of the Super-Being (although even in early Soviet Marxism "God-building" had a sub-rosa tradition). In Turchin's case we are probably justified in linking the inclusion of Hegelian concepts in his interpretation of nature to the education in philosophy he received in the Soviet Union. Soviet Marxism was probably one of several sources of Turchin's philosophic views; others are cybernetics and the thought of such earlier writers on cosmic evolution as Chardin and Vernadsky.

In view of the links one can see between the ideas of Turchin and Marxism, it is particularly interesting to notice that Turchin is now in political difficulty in the Soviet Union. Before I give some of the details of his political biography, however, I shall note that in this essentially nonpolitical manuscript Turchin gives a few hints of possible social implications of his interpretation. He remarks that the cybernetic view he is presenting places great emphasis on "control" and that it draws an analogy between society and a multicellular organism. He then observes, "This point of view conceals in itself a great danger that in vulgarized form it can easily lead to the conception of a fascist-type totalitarian state." This possibility of a totalitarian state, of whatever type, is clearly repugnant to Turchin, and his personal experience is a witness that he is willing to risk his own security in order to struggle against such a state. As for his interpretation of social evolution, he contends that "the possibility that a theory can be vulgarized is in no way an argument against its truth." In the last sections of his book he presents suggestions for avoiding such vulgarizations while still working for greater social integration.

Turchin is wrestling in this last part of his interpretation with a problem that has recently plagued many thinkers in Western Europe and America as well: Can one combine a scientific explanation of man and society with a commitment to individual freedom and social justice? Turchin is convinced that such combination of goals is possible; indeed, he sees this alliance as imperative, since he believes there is no conceptual alternative to the scientific worldview and no ethical alternative to the maintenance of individual freedom. It is the steadfastness of his support of science that will seem surprising to some of his readers in the West, where science is often seen as only a partial worldview, one to be supplemented with religious or nonscientific ethical or esthetic principles. Turchin, however, believes that humans can be explained within an entirely naturalistic framework.

His belief that ethical and altruistic modes of behavior can emerge from an evolutionary scheme is, therefore, one that brings him in contact with recent writers in the West on sociobiology, physical anthropology, and evolutionary behavior. His emphases on information theory, on irreducible levels, and on the dangers of vulgarizations of scientific explanations of human behavior while nonetheless remaining loyal to science may make contributions to these already interesting discussions.

———————◆•••◆———————

Valentin Fedorovich Turchin, born in 1931, holds a doctor's degree in the physical and mathematical sciences. He worked in the Soviet science center in Obninsk, near Moscow, in the Physics and Energetics Institute and then later became a senior scientific researcher in the Institute of Applied Mathematics of the Academy of Sciences of the USSR. In this institute he specialized in information theory and the computer sciences. While working in these fields he developed a new computer language that was widely applied in the USSR, the "Refal" system. After 1973 he was the director of a laboratory in the Central Scientific-Research Institute for the Design of Automated Construction Systems. During his years of professional employment Dr. Turchin published over 65 works in his field. In sum, in the 1960s and early 1970s, Valentin Turchin was considered one of the leading computer specialists in the Soviet Union.

Dr. Turchin's political difficulties began in 1968, when he was one of hundreds of scientists and other liberal intellectuals who signed letters protesting the crackdown on dissidents in the Soviet Union preceding and accompanying the Soviet-led invasion of Czechoslovakia. In the same year he wrote an article entitled ''The Inertia of Fear'' which circulated widely in *samizdat,* the system of underground transmission of manuscripts in the Soviet Union. Later the same article was expanded into a book-length manuscript in which Dr. Turchin criticized the vestiges of Stalinism in Soviet society and called for democratic reform.

In September 1973 Dr. Turchin was one of the few people in the Soviet Union who came to the defense of the prominent Soviet physicist Andrei D. Sakharov when the dissident scientist was attacked in the Soviet press. As a result of his defense of Sakharov, Turchin was denounced in his institute and demoted from chief of laboratory to senior research associate. The computer scientist continued his defense of human rights, and in July 1974, he was dismissed from the institute. In the ensuing months Dr. Turchin found that he had been blacklisted at other places of employment.

In the last few years Professor Turchin has been chairman of the Moscow chapter of Amnesty International, an organization that has worked for human rights throughout the world. When other Soviet scholars were persecuted, including Andrei Tverdokhlebov and Sergei Kovalev, Dr. Turchin helped publicize their plight. During this period, his wife, a mathematician, has financially supported her husband and their two sons.

In 1974 and 1975 Dr. Turchin received invitations to teach at several American universities, but the Soviet government refused to grant him an exit visa. Several writers in the West speculated that he would soon be arrested and tried, but so far he has been able to continue his activity, working within necessary limits. His apartment has been searched by the police and he has been interrogated.

Dr. Turchin wrote *The Phenomenon of Science* before these personal difficulties began, and he did not intend it to be a political statement. Indeed, the manuscript was accepted for publication by a leading Soviet publishing house, and preliminary Soviet reviewers praised its quality. Publication of the book was stopped only after Dr. Turchin was criticized on other grounds. Therefore, that the initial publication of *The Phenomenon of Science* is outside the Soviet Union, should not be seen as a result of its content, but of the non-scientific activities of its author after it was written.

LOREN R. GRAHAM

Columbia University
June 1977

Preface

WHAT IS scientific knowledge of reality? To answer this question from a scientific point of view means to look at the human race from outside, from outer space so to speak. Then human beings will appear as certain combinations of matter which perform certain actions, in particular producing some kind of words and writing some kind of symbols. How do these actions arise in the process of life's evolution? Can their appearance be explained on the basis of some general principles related to the evolutionary process? What is scientific activity in light of these general principles? These are the questions we shall attempt to answer in this book.

Principles so general that they are applicable both to the evolution of science and to biological evolution require equally general concepts for their expression. Such concepts are offered by cybernetics, the science of relationships, control, and organization in all types of objects. Cybernetic concepts describe physicochemical, biological, and social phenomena with equal success. It is in fact the development of cybernetics, and particularly its successes in describing and modeling purposeful behavior and in pattern recognition, which has made the writing of this book possible. Therefore it would be more precise to define our subject as the cybernetic approach to science as an object of study.

The intellectual pivot of the book is the concept of the *metasystem transition*—the transition from a cybernetic system to a metasystem, which includes a set of systems of the initial type organized and controlled in a definite manner. I first made this concept the basis of an analysis of the development of sign systems used by science. Then, however, it turned out that investigating the entire process of life's evolution on earth from this point of view permits the construction of a coherent picture governed by uniform laws. Actually it would be better to say a moving picture, one which begins with the first living cells and ends with present-day scientific theories and the system of industrial production. This moving picture shows, in particular, the place of the phenomenon of science among the other phenomena of the world and reveals the significance of science in the overall picture of the evolution of the universe. That is how the plan of this book arose. How convincingly this picture has been drawn I propose to leave to the reader's judgment.

In accordance with the plan of the book, many very diverse facts and conceptions are presented. Some of the facts are commonly known; I try to limit my discussion of them, fitting them into the system and relating them to my basic idea. Other facts are less well-known, and in such cases I dwell on them in more detail. The same is true for the conceptions; some are commonly recognized while others are less well known and, possibly, debatable. The varied nature of the material creates a situation where different parts of the book require different efforts from the reader. Some parts are descriptive and easy to read, in other places it is necessary to go deeply into quite specialized matters. Because the book is intended for a broad range of readers and does not assume knowledge beyond the secondary school level, I provide the necessary theoretical information in all such cases. These pages will require a certain effort of the untrained reader.

The book gives an important place to the problems of the theory of knowledge and logic. They are, of course, treated from a cybernetic point of view. Cybernetics is now waging an attack on traditional philosophical epistemology, offering a new natural-science interpretation of some of its concepts and rejecting others as untenable. Some philosophers oppose the rise of cybernetics and consider it an infringement on their territory. They accuse cyberneticists of making the truth "crude" and "simplifying" it; they claim cyberneticists ignore the "fundamental difference" between different forms of the movement of matter (and this is despite the thesis of the world's unity!). But the philosopher to whom the possessive attitude toward various fields of knowledge is foreign should welcome the attacks of the cyberneticists. The development of physics and astronomy once destroyed natural philosophy, sparing philosophers of the need to talk approximately about a subject which scientists could discuss exactly. It appears that the development of cybernetics will do the same thing with philosophical epistemology or, to be more cautious, with a significant part of it. This should be nothing but gratifying. Philosophers will always have enough concerns of their own; science rids them of some, but gives them others.

Because the book is devoted to science in toto as a definite method of interaction between human society and its environment, it contains practically no discussion of concrete natural-science disciplines. The presentation remains entirely at the level of the concepts of cybernetics, logic, and mathematics, which are equally significant for all modern science. The only exception is for some notions of modern physics which are fundamentally important for the theory of sign systems. A concrete analysis of science's interaction with production and social life was also outside the scope of the problem. This is a distinct matter to which a vast literature has been devoted; in this book I remain at the level of general cybernetic concepts.

It is dangerous to attempt to combine a large amount of material from different fields of knowledge into a single, whole picture; details may become distorted, for a person cannot be a specialist in everything. Because this book attempts precisely to create such a picture, it is very likely that specialists in the fields of science touched on here will find omissions and inaccuracies; such is the price which must be paid for a wide scope. But such pictures are essential. It only remains for me to hope that this book contains nothing more than errors in detail which can be eliminated without detriment to the overall picture.

V. F. TURCHIN

THE
PHENOMENON
OF SCIENCE

CHAPTER ONE
The Initial Stages of Evolution

■ THE BASIC LAW OF EVOLUTION

IN THE PROCESS of the evolution of life, as far as we know, the total mass of living matter has always been and is now increasing and growing more complex in its organization. To increase the complexity of the organization of biological forms, nature operates by trial and error. Existing forms are reproduced in many copies, but these are not identical to the original. Instead they differ from it by the presence of small random variations. These copies then serve as the material for natural selection. They may act as individual living beings, in which case selection leads to the consolidation of useful variations, or elements of more complex forms, in which case selection is also directed to the structure of the new form (for example, with the appearance of multicellular organisms). In both cases selection is the result of the struggle for existence, in which more viable forms supplant less viable ones.

This mechanism of the development of life, which was discovered by Charles Darwin, may be called the basic law of evolution. It is not among our purposes to substantiate or discuss this law from the point of view of those laws of nature which could be declared more fundamental. We shall take the basic law of evolution as given.

■ THE CHEMICAL ERA

THE HISTORY OF LIFE before the appearance of the human being can be broken into two periods, which we shall call the "chemical" era and the "cybernetic" era. The bridge between them is the emergence of animals with distinct nervous systems, including sense organs, nerve fibers for transmitting information, and nerve centers (nodes) for converting this information. Of course, these two terms do not signify that the concepts and methods of cybernetics are inapplicable to life in the "chemical" era; it is simply that the animal of the "cybernetic" era is the classical object of cybernetics, the one to which its appearance and establishment as a scientific discipline are tied.

We shall review the history and logic of evolution in the precybernetic period only in passing, making reference to the viewpoints of present-day biologists.[1] Three stages can be identified in this period.

In the first stage the chemical foundations of life are laid. Macromolecules of nucleic acids and proteins form with the property of replication, making copies or "prints" where one macromolecule serves as a matrix for synethesizing a similar macromolecule from elementary radicals. The basic law of evolution, which comes into play at this stage, causes matrices which have greater reproductive intensity to gain an advantage over matrices with lesser reproductive intensity, and as a result more complex and active macromolecules and systems of macromolecules form. Biosynthesis demands free energy. Its primary source is solar radiation. The products of the partial decay of life forms that make direct use of solar energy (photosynthesis) also contain a certain reserve of free energy which may be used by the already available chemistry of the macromolecule. Therefore, this reserve is used by special forms for which the products of decay serve as a secondary source of free energy. Thus the division of life into the plant and animal worlds arises.

[1] I am generally following the report by S. E. Schnoll entitled "The Essence of Life. Invariance in the General Direction of Biological Evolution," in *Materialy seminara "Dialektika i sovremennoe estestvoznanie"* (Materials of the "Dialectics and Modern Natural Science" Seminar), Dubna, 1967.

2   INITIAL STAGES OF EVOLUTION

The second stage of evolution is the appearance and development of the motor apparatus in animals.

Plants and animals differ fundamentally in the way they obtain energy. With a given level of illumination the intensity of absorption of solar energy depends entirely on the amount of plant surface, not on whether it moves or remains stationary. Plants were refined by the creation of outlying light catchers—green leaves secured to a system of supports and couplings (stems, branches, and the like). This design works very well, ensuring a slow shift in the green surfaces toward the light which matches the slow change in illumination.

The situation is entirely different with animals, in particular with the most primitive types such as the amoeba. The source of energy—food—fills the environment around it. The intake of energy is determined by the speed at which food molecules are diffused through the shell that separates the digestive apparatus from the external environment. The speed of diffusion depends less on the size of the surface of the digestive apparatus than on the movement of this surface relative to the environment; therefore it is possible for the animal to take in food from different sectors of the environment. Consequently, even simple, chaotic movement in the environment or, on the other hand, movement of the environment relative to the organism (as is done, for example, by sponges which force water through themselves by means of their cilia) is very important for the primitive animal and, consequently, appears in the process of evolution. Special forms emerge (intracellular formations in one-celled organisms and ones containing groups of cells in multicellular organisms) whose basic function is to produce movement.

In the third stage of evolution the movements of animals become directed and the incipient forms of sense organs and nervous systems appear in them. This is also a natural consequence of the basic law. It is more advantageous for the animal to move in the direction where more food is concentrated, and in order for it to do so it must have sensors that describe the state of the external environment in all directions (sense organs) and information channels for communication between these sensors and the motor apparatus (nervous system). At first the nervous system is extremely primitive. Sense organs merely distinguish a few situations to which the animal must respond differently. The volume of information transmitted by the nervous system is slight and there is no special apparatus for processing the information. During the process of evolution the sense organs become more complex and deliver an increasing amount of information about the external environment. At the same time the motor apparatus is refined, which makes ever-increasing demands on the carrying capacity of the nervous system. Special formations appear—nerve centers which convert information received from the sense organs into information controlling the organs of movement. A new era begins: the "cybernetic" era.

■ CYBERNETICS

TO ANALYZE evolution in the cybernetic period and to discover the laws governing the organization of living beings in this period (for brevity we will call them "cybernetic animals") we must introduce certain fundamental concepts and laws from cybernetics.

The term "cybernetics" itself was, of course, introduced by Norbert Wiener, who defined it descriptively as the theory of relationships and control in the living organism and the machine. As is true in any scientific discipline, a more precise definition of cybernetics requires the introduction of its basic concepts. Properly speaking, to introduce the basic concepts is the same as defining a particular science, for all that remains to be added is that a description of the world by means of this system of concepts is, in fact, the particular, concrete science.

Cybernetics is based above all on the concept of the *system,* a certain material object which consists of other objects which are called *subsystems* of the given system. The subsystem of a certain system may, in its turn, be viewed as a system consisting of other subsystems. To be precise, therefore, the meaning of the concept we have introduced does not lie in the term "system" by itself, that is, not in ascribing the property of "being a system" to a certain object (this is quite meaningless, for any object may be considered a system), but rather in the connection between the terms "system" and "subsystem," which reflects a definite relationship among objects.

The second crucial concept of cybernetics is the concept of the *state* of a system (or subsystem). Just as the concept of the system relies directly on our spatial intuition, the concept of state relies directly on our intuition of time and it canot be defined except by referring to experience. When we say that an object has changed in some respect we are saying that it has passed into a different state. Like the concept of system, the concept of state is a concealed relationship: the relationship between two moments in time. If the world were immobile the concept of state would not occur, and in those disciplines where the world is viewed statically, for example in geometry, there is no concept of state.

Cybernetics studies the organization of systems in space and time, that is, it studies how subsystems are connected into a system and how change in the state of some subsystems influences the state of other subsystems. The primary emphasis, of course, is on organization in time which, when it is purposeful, is called *control*. Causal relations between states of a system and the characteristics of its behavior in time which follow from this are often called the *dynamics* of the system, borrowing a term from physics. This term is not applicable to cybernetics, because when we speak of the dynamics of a system we are inclined to view it as something whole, whereas cybernetics is concerned mainly with investigating the mutual influences of subsystems making up the particular system. Therefore, we prefer to speak of *organization in time,* using the term *dynamic* description only when it must be juxtaposed to the *static* description which considers nothing but spatial relationships among subsystems.

A cybernetic description may have different levels of detail. The same system may be described in general outline, in which it is broken down into a few large subsystems or "blocks," or in greater detail, in which the structure and internal connections of each block are described. But there is always some final level beyond which the cybernetic description does not apply. The subsystems of this level are viewed as elementary and incapable of being broken down into constituent parts. The real physical nature of the elementary subsystems is of no interest to the cyberneticist, who is concerned only with how they are interconnected. The nature of two physical objects may be radically different, but if at some level of cybernetic description they are organized from subsystems in the same way (considering the dynamic aspect!), then from the point of view of cybernetics they can be considered, at the given level of description, identical. Therefore, the same cybernetic considerations can be applied to such different objects as a radar circuit, a computer program, or the human nervous system.

■ DISCRETE AND CONTINUOUS SYSTEMS

THE STATE OF A SYSTEM is defined through the aggregate of states of all its subsystems, which in the last analysis means the elementary subsystems. There are two types of elementary subsystems: those with a finite number of possible states, also called subsystems with discrete states, and those with an infinite number, also called subsystems with continuous states. The wheel of a mechanical calculator or taxi meter is an example of a subsystem with discrete states. This wheel is normally in one of 10 positions which correspond to the 10 digits between 0 and 9. From time to time it turns and passes from one state into another. This process of turning does not interest us. The correct functioning of the system (of the calculator or meter) depends entirely on how the "normal" positions of the wheels are interconnected, while how the change from one position (state) to another takes place is inconsequential. Therefore we can consider the calculator as a system whose elementary subsystems can only be in discrete states. A modern high-speed digital computer also consists of subsystems (trigger circuits) with discrete states. Everything that we know at the present time regarding the nervous systems of humans and animals indicates that the interaction of subsystems (neurons) with discrete states is decisive in their functioning.

On the other hand, a person riding a bicycle and an analog computer are both examples of systems consisting of subsystems with continuous states. In the case of the bicycle rider these subsystems are all the parts of the bicycle and human body which are moving relative to one another: the wheels, pedals, handlebar, legs, arms, and so on. Their states are their positions in space. These positions are described by coordinates (numbers) which can assume continuous sets of values.

If a system consists exclusively of subsystems with discrete states then the system as a whole must be a system with discrete states. We shall simply call such systems "discrete systems," and we shall call systems with continuous sets of states "continuous systems." In many respects discrete systems are simpler to analyze than continuous ones. Counting the number of possible states of a system, which plays an important part in cybernetics, requires only a knowledge of elementary arithmetic in the case of discrete systems. Suppose discrete system A consists of two subsystems a_1 and a_2; subsystem a_1 may have n_1 possible states, while subsystem a_2 may have n_2. Assuming that each state of system a_1 can combine with each state of system a_2, we find that N, the number of possible states of system A, is $n_1 n_2$. If system A consists of m subsystems a_i where $i = 1, 2, \ldots ,$ m, then

$$N = n_1 n_2 \ldots n_m$$

From this point on we shall consider only discrete systems. In addition to the pragmatic consideration that they are simpler in principle than continuous systems, there are two other arguments for such a restriction.

First, all continuous systems can in principle be viewed as discrete systems with an extremely large number of states. In light of the knowledge quantum physics has given us, this approach can even be considered theoretically more correct. The reason why continuous systems do not simply disappear from cybernetics is the existence of a very highly refined apparatus for consideration of such systems: mathematical analysis, above all, differential equations.

Second, the most complex cybernetic systems, both those which have arisen naturally and those created by human hands, have invariably proved to be discrete. This is seen especially clearly in the example of animals. The relatively simple biochemical mechanisms that regulate body temperature, the content of various substances in the blood, and similar characteristics are continuous, but the nervous system is constructed according to the discrete principle.

■ THE RELIABILITY OF DISCRETE SYSTEMS

WHY DO DISCRETE SYSTEMS prove to be preferable to continuous ones when it is necessary to perform complex functions? Because they have a much higher reliability. In a cybernetic device based on the principle of discrete states each elementary subsystem may be in only a small number of possible states, and therefore the system ordinarily ignores small deviations from the norm of various physical parameters of the system, reestablishing one of its permissible states in its "primeval purity." In a continuous system, however, small disturbances continuously accumulate and if the system is too complex it ceases functioning correctly. Of course, in the discrete system too there is always the possibility of a breakdown, because small changes in physical parameters do lead to a finite probability that the system will switch to an "incorrect" state. Nonetheless, discrete systems definitely have the advantage. Let us demonstrate this with the following simple example.

Suppose we must transmit a message by means of electric wire over a distance of, say, 100 kilometers (62 miles). Suppose also that we are able to set up an automatic station for every kilometer of wire and that this station will amplify the signal to the power it had at the previous station and, if necessary, convert the signal (see figure 1.1). We assume that the maximum signal our equipment permits us to send has a magnitude of one volt and that the average distortion of the signal during transmission from station to station (noise) is equal to 0.1 volt.

First let us consider the continuous method of data transmission. The content of the message will be the amount of voltage applied to the wire at its beginning. Owing to noise, the voltage at the other end of the wire—the message received—will differ from the initial voltage. How great will this difference be? Considering noise in different segments of the line to be independent, we find that after the signal

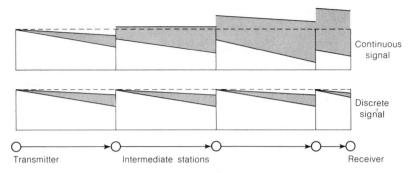

Figure 1.1. Transmission of a signal in continuous and discrete systems. (The shaded part shows the area of signal ambiguity.)

passes the 100 stations the root-mean square magnitude of noise will be one volt (the mean squares of noise are summed). Thus, average noise is equal to the maximum signal, and it is therefore plain that we shall not in fact receive any useful information. Only by accident can the value of the voltage received coincide with the value of the voltage transmitted. For example, if a precision of 0.1 volt satisfies us the probability of such a coincidence is approximately 1/10.

Now let us look at the discrete variant. We shall define two "meaningful" states of the initial segment of the wire: when the voltage applied is equal to zero and when it is maximal (one volt). At the intermediate stations we install automatic devices which transmit zero voltage on if the voltage received is less than 0.5 volt and transmit a normal one-volt signal if the voltage received is more than 0.5 volt. In this case, therefore, for one occasion (one signal) information of the "yes/no" type is transmitted (in cybernetics this volume of information is called one "bit"). The probability of receiving incorrect information depends strongly on the law of probability distribution for the magnitude of noise. Noise ordinarily follows the so-called normal law. Assuming this law we can find that the probability of error in transmission from one station to the next (which is equal to the probability that noise will exceed 0.5 volt) is $0.25 \cdot 10^{-6}$. Thus the probability of an error in transmission over the full length of the line is $0.25 \cdot 10^{-4}$. To transmit the same message as was transmitted in the previous case—that is, a value between 0 and 1 with a precision of 0.1 of a certain quantity lying between 0 and 1—all we have to do is send four "yes/no" type signals. The probability that there will be error in at least one of the signals is 10^{-4}. Thus, with the discrete method the total probability of error is 0.01 percent, as against 90 percent for the continuous method.

■ INFORMATION

WHEN WE BEGAN describing a concrete cybernetic system it was impossible not to use the term *information*—a word familiar and understandable in its informal, conversational meaning. The cybernetic concept of information, however, has an exact quantitative meaning.

Let us imagine two subsystems A and B (see figure 1.2), which are interconnected in such a way that a change in the state of A leads to a change in the state of B. This can also be expressed as follows: A influences B. Let us consider the state of B at a certain moment in time t_1 and at a later moment t_2. We shall signify the first state as S_1 and the second as S_2. State S_2 depends on state S_1. The relationship of S_2 to S_1 is probabilistic, however, not unique. This is because we are not considering an idealized theoretical system governed by a deterministic law of movement but rather a real system whose states S_i are the results of experimental data. With such an approach we may also speak of the law of movement, understanding it in the probabilistic sense—that is, as the conditional probability of state S_2 at moment t_2 on the condition that the system was in state S_1 at moment t_1. Now let us momentarily ignore the law of movement. We shall use N to designate the total number of possible states of subsystem B and imagine that conditions are such that at any moment in time system B can as-

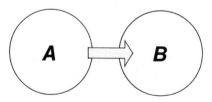

Figure 1.2.

sume any of N states with equal probability, regardless of its state at the preceding moment. Let us attempt to give a quantitative expression to the degree (or strength) of the cause–effect influence of system A on such an inertialess and "lawless" subsystem B. Suppose B acted upon by A switches to a certain completely determinate state. It is clear that the "strength of influence" which is required from A for this depends on N, and will be larger as N is larger. For example, if $N = 2$ then B, even if it is completely unrelated to A, when acted upon by random factors can switch with a probability of .5 to the very state A "recommends." But if $N = 10^9$, when we have noticed such a coincidence we shall hardly doubt the influence of A on B. Therefore, some monotonic increasing function of N should serve as the measure of the "strength of the influence" of A on B. What this essentially means is that it serves as a measure of the intensity of the cause–effect relationship between two events, the state of A in the time interval from t_1 to t_2 and the state of B at t_2. In cybernetics this measure is called the quantity of information transmitted from A to B between moments in time t_1 and t_2, and a logarithm serves as the monotonic increasing function. So, in our example, the quantity of information I passed from A to B is equal to log N.

Selection of the logarithmic function is determined by its property according to which

$$\log N_1 N_2 = \log N_1 + \log N_2$$

Suppose system A influences system B, which consists of two independent subsystems B_1 and B_2 with number of possible states N_1 and N_2 respectively (see figure 1.3). Then the number of states of system B is $N_1 N_2$ and the quantity of information I that must be transmitted to system B in order for it to assume one definite state is, owing to the above-indicated property of the logarithm, the sum

$$I = \log N_1 N_2 = \log N_1 + \log N_2 = I_1 + I_2$$

where I_1 and I_2 are the quantities of information required by subsystems B_1 and B_2. Thanks to this property the information assumes definite characteristics of a substance; it spreads over the independent subsystems like a fluid filling a number of vessels. We are speaking

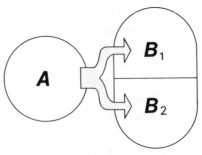

Figure 1.3.

of the joining and separation of information flows, information capacity, and information processing and storage.

The question of information storage is related to the question of the law of movement. Above we mentally set aside the law of movement in order to define the concept of information transmission. If we now consider the law of movement from this new point of view, it can be reduced to the transmission of information from system B at moment t_1 to the same system B at moment t_2. If the state of the system does not change with the passage of time, this is information storage. If state S_2 is uniquely determined by S_1 at a preceding moment in time the system is called *fully deterministic*. If S_1 is uniquely determined by S_2 the system is called *reversible;* for a reversible system it is possible in principle to compute all preceding states on the basis of a given state because information loss does not occur. If the system is not reversible information is lost. The law of movement is essentially something which regulates the flow of information in time from the system and back to itself.

Figure 1.4 shows the chart of information transmission from system A to system C through system B. B is called the *communication channel*. The state of B can be influenced not only by the state of system A, but also by a certain uncontrolled factor X, which is called *noise*. The final state of system C in this case depends not only on the state of A, but also on factor X (information distortion). One more important diagram of information exchange is shown in figure 1.5. This is the so-called *feedback* diagram. The state of system A at t_1

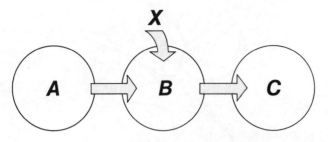

Figure 1.4.

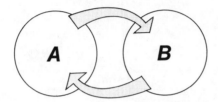

Figure 1.5.

influences the state of B at t_2, then the latter influences the state of A at t_3. The circle of information movement is completed.

With this we conclude for now our familiarization with the general concepts of cybernetics and turn to the evolution of life on earth.

■ THE NEURON

THE EXTERNAL APPEARANCE of a nerve cell (neuron) is shown schematically in figure 1.6. A neuron consists of a fairly large (up to 0.1 mm) cell body from which several processes called *dendrites* spread, giving rise to finer and finer processes like the branching of a tree. In addition to the dendrites one other process branches out from the body of the nerve cell. This is the *axon,* which resembles a long, thin wire. Axons can be very long, up to a meter, and they end in tree-like branching systems as do the dendrites. At the ends of the branches coming from the axon one can see small plates or bulblets. The bulblets of one neuron approach close to different segments of the body or dendrites of another neuron, almost touching them. These

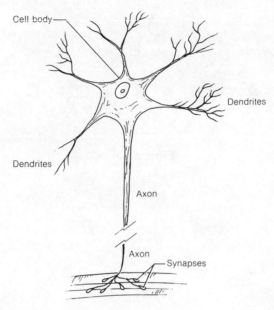

Figure 1.6. Diagram of the structure of a neuron.

contacts are called *synapses,* and it is through them that neurons interact with one another. The number of bulblets approaching the dendrites of the single neuron may run into the dozens and even hundreds. In this way the neurons are closely interconnected and form a *nerve net.*

When one considers certain physicochemical properties (above all the propagation of electrical potential over the surface of the cell) one discovers the neurons can be in one of two states—the state of dormancy or the state of stimulation. From time to time, influenced by other neurons or outside factors, the neuron switches from one state to the other. This process takes a certain time, of course, so that an investigator who is studying the dynamics of the electrical state of a neuron, for example, considers it a system with continuous states. But the information we now have indicates that what is essential for the functioning of the nervous system as a whole is not the nature of switching processes but the very fact that the particular neurons are in one of these two states. Therefore, we may consider that the nerve net is a discrete system which consists of elementary subsystems (the neurons) with two states.

When the neuron is stimulated, a wave of electrical potential runs along the axon and reaches the bulblets in its branched tips. From the bulblets the stimulation is passed across the synapses to the corresponding sectors of the cell surface of other neurons. The behavior of a neuron depends on the state of its synapses. The simplest model of the functioning of the nerve net begins with the assumption that the state of the neuron at each moment in time is a single-valued function of the state of its synapses. It has been established experimentally that the stimulation of some synapses promotes stimulation of the cell, whereas the stimulation of other synapses prevents stimulation of the cell. Finally, certain synapses are completely unable to conduct stimulation from the bulblets and therefore do not influence the state of the neuron. It has also been established that the conductivity of a synapse increases after the first passage of a stimulus through it. Essentially a closing of the contact occurs. This explains how the system of communication among neurons, and consequently the nature of the nerve net's functioning, can change without a change in the relative positions of the neurons.

The idea of the neuron as an instantaneous processor of information received from the synapses is, of course, very simplified. Like any cell the neuron is a complex machine whose functioning has not yet been well understood. This machine has a large internal memory, and therefore its reactions to external stimuli may show great variety. To understand the general rules of the working of the nervous system, however, we can abstract from these complexities (and really, we have no other way to go!) and begin with the simple model outlined above.

■ THE NERVE NET

A GENERALIZED DIAGRAM of the nerve system of the "cybernetic animal" in its interaction with the environment is shown in figure 1.7. Those sensory nerve cells which are stimulated by the action of outside factors are called *receptors* (that is, receivers) because they

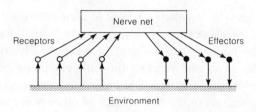

Figure 1.7. Nervous system of the "cybernetic animal."

are the first to receive information about the state of the environment. This information enters the nerve net and is converted by it. As a result certain nerve cells called *effectors* are stimulated. Branches of the effector cells penetrate those tissues of the organism which the nervous system affects directly. Stimulation of the effector causes a contraction of the corresponding muscle or the stimulation of the activity of the appropriate gland. We shall call the state of all receptors at a certain moment in time the *situation* at that moment. (It would be more precise—if more cumbersome—to say the "result of the effect of the situation on the sense organs.") We will call the state of all the effectors the *"action."* Therefore, the role of the nerve net is to convert a *situation* into an *action*.

It is convenient to take the term "environment" from figure 1.7 to mean not just the objects which surround the animal, but also its bone and muscle system and generally everything that is not part of the nervous system. This makes it unnecessary to give separate representations in the diagram to the animal body and what is not the body, especially because this distinction is not important in principle for the activity of the nervous system. The only thing that is important is that stimulation of the effectors leads to certain changes in the "environment." With this general approach to the problem as the basis of our consideration, we need only classify these changes as "useful" or "harmful" for the animal without going into further detail.

The objective of the nervous system is to promote the survival and reproduction of the animal. The nervous system works well when stimulation of the effectors leads to changes in the state of the environment that help the animal survive or reproduce, and it works badly when it leads to the reverse. With its increasing refinement in the process of evolution, the nervous system has performed this task increasingly well. How does it succeed in this? What laws does this process of refinement follow?

We will try to answer these questions by identifying in the evolution of the animal nervous system several stages that are clearly distinct from a cybernetic point of view and by showing that the transition from each preceding stage to each subsequent stage follows inevitably from the basic law of evolution. Because the evolution of living beings in the cybernetic era primarily concerns the evolution of their nervous systems, a periodization of the development of the nervous system yields a periodization of the development of life as a whole.

■ THE SIMPLE REFLEX (IRRITABILITY)

THE SIMPLEST VARIANT of the nerve net is when there is no net at all. In this case the receptors are directly connected to the effectors and stimulation from one or several receptors is transmitted to one or several effectors. We shall call such a direct connection between stimulation of a receptor and an effector the *simple reflex.*

This stage, the third in our all-inclusive enumeration of the stages of evolution, is the bridge between the chemical and cybernetic eras. The Coelenterata are animals fixed at the level of the simple reflex. As an example let us take the hydra, which is studied in school as a typical representative of the Coelenterata. The body of a hydra has the shape of an elongated sac. Its interior, the coelenteron, is connected to the environment through a mouth, which is surrounded by several tentacles. The walls of the sac consist of two layers of cells: the inner layer (entoderm) and the outer layer (ectoderm). Both the ectoderm and the entoderm have many muscle cells which contain small fibers that are able to contract, thus setting the body of the hydra in motion. In addition, there are nerve cells in the ectoderm; the cells located closest to the surface are receptors and the cells which are set deeper, among the muscles, are effectors. If a hydra is pricked with a needle it squeezes itself into a tiny ball. This is a simple reflex caused by transmission of the stimulation from the receptors to the effectors.

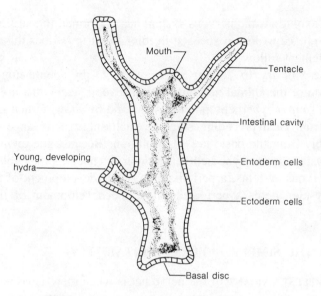

Figure 1.8. The structure of the hydra.

But the hydra is also capable of much more complex behavior. After it has captured prey, the hydra uses its tentacles to draw the prey to its mouth and then swallows the prey. This behavior can also be explained by the aggregate action of simple reflexes connecting effectors and receptors locally, within small segments of the body. For example, the model of a tentacle (depicted in figure 1.9) explains its

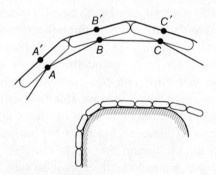

Figure 1.9. Model of a tentacle.

ability to wrap itself around captured objects. Let us picture a certain number of links connected by hinges (for simplicity we shall consider a two-dimensional picture). Points A and B, A' and B', B and C, and B' and C', etc. are interconnected by strands which can contract (muscles). All these points are sensitive and become stimulated when they touch an object (receptors). The stimulation of each point causes a contraction of the two strands connected to it (reflex).

■ THE COMPLEX REFLEX

THE SIMPLE REFLEX relationship between the stimulated cell and the muscle cell arises naturally, by the trial and error method, in the process of evolution. If the correlation between stimulation of one cell and contraction of another proves useful for the animal, then this correlation becomes established and reinforced. Where interconnected cells are mechanically copied in the process of growth and reproduction, nature receives a system of parallel-acting simple reflexes resembling the tentacle of the hydra. But when nature has available a large number of receptors and effectors which are interconnected by pairs or locally, there is a "temptation" to make the system of connections more complex by introducing intermediate neurons. This is advantageous because where there is a system of connections among all neurons, forms of behavior that are not possible where all connections are limited to pairs or localities now become so. This point can be demonstrated by a simple calculation of all the possible methods of converting a *situation* into an *action* with each method of interconnection. For example, assume that we have n receptors and effectors connected by pairs. In each pair the connection may be positive (stimulation causes stimulation and dormancy evokes dormancy) or negative (stimulation evokes dormancy and dormancy causes stimulation). In all, therefore, 2^n variants are possible, which means 2^n variants of behavior. But if we assume that the system of connections can be of any kind, which is to say that the state of each effector (stimulation or dormancy) can depend in any fashion on the state of all the receptors, then a calculation of all possible variants of behavior yields the number $2^{(2^n)n}$, which is immeasurably larger than 2^n.

Exactly the same calculation leads to the conclusion that joining any subsystems which join independent groups of receptors and effectors into a single system always leads to an enormous increase in the number of possible variants of behavior. Throughout the entire course of the history of life, therefore, the evolution of the nervous system has progressed under the banner of increasing centralization.

But "centralization" can mean different things. If all neurons are connected in one senselessly confused clump, then the system—despite its extremely "centralized" nature—will hardly have a chance to survive in the struggle for existence. Centralization poses the following problem: how to select from all the conceivable ways of joining many receptors with many effectors (by means of intermediate neurons if necessary) that way which will correlate a correct action (that is, one useful for survival and reproduction) to each situation? After all, a large majority of the ways of interconnection do not have this characteristic.

We know that nature takes every new step toward greater complexity in living structures by the trial and error method. Let us see what direct application of the trial and error method to our problem yields. As an example we shall consider a small system consisting of 100 receptors and 100 effectors. We shall assume that we have available as many neurons as needed to create an intermediate nerve net and that we are able to determine easily whether the particular method of connecting neurons produces a correct reaction to each situation. We shall go through all conceivable ways of connection until we find the one we need. Where $n = 100$ the number of functionally different nerve nets among n receptors and n effectors is

$$2^{(2^n)n} \approx 10^{(10^{32})}$$

This is an inconceivably large number. We cannot sort through such a number of variants and neither can Mother Nature. If every atom in the entire visible universe were engaged in examining variants and sorting them at a speed of 1 billion items a second, even after billions of billions of years (and our earth has not existed for more than 10 billion years) not even one billionth of the total number of variants would have been examined.

But somehow an effectively functioning nerve net does form! And higher animals have not hundreds or thousands but millions of receptors and effectors. The answer to the riddle is concealed in the *hierarchical structure* of the nervous system. Here again we must make an excursion into the area of general cybernetic concepts. We shall call the fourth stage of evolution the stage of the *complex reflex,* but we shall not be able to define this concept until we have familiarized ourselves with certain facts about hierarchically organized nerve nets.

CHAPTER TWO
Hierarchical Structures

■ THE CONCEPT OF THE CONCEPT

LET US LOOK at a nerve net which has many receptors at the input but just one effector at the output. Thus, the nerve net divides the set of all situations into two subsets: situations that cause stimulation of the effector and situations that leave it dormant. The task being performed by the nerve net in this case is called *recognition* (discrimination), recognizing that the situation belongs to one of the two sets. In the struggle for existence the animal is constantly solving recognition problems, for example, distinguishing a dangerous situation from one that is not, or distinguishing edible objects from inedible ones. These are only the clearest examples. A detailed analysis of animal behavior leads to the conclusion that the performance of any complex action requires that the animal resolve a large number of ''small'' recognition problems continuously.

In cybernetics a set of situations is called a *concept*.[1] To make clear how the cybernetic understanding of the word ''concept'' is related to its ordinary meaning let us assume that the receptors of the nerve net under consideration are the light-sensitive nerve endings of the retina of the eye or, speaking in general, some light-sensitive points on a screen which feed information to the nerve net. The

---

[1] Later we shall give a somewhat more general definition of the concept and a set of situations shall be called an Aristotelian concept. At present we shall drop the adjective ''Aristotelian'' for brevity.

receptor is stimulated when the corresponding sector of the screen is illuminated (more precisely, when its illumination is greater than a certain threshold magnitude) and remains dormant if the sector is not illuminated. If we imagine a light spot in place of each stimulated receptor and a dark spot in place of each unstimulated one, we shall obtain a picture that differs from the image striking the screen only by its discrete nature (the fact that it is broken into separate points) and by the absence of semitones. We shall consider that there are a large number of points (receptors) on the screen and that the images which can appear on the screen ("pictures") have maximum contrasts—that is, they consist entirely of black and white. Then each situation corresponds to a definite picture.

According to traditional Aristotelian logic, when we think or talk about a definite picture (for example the one in the upper left corner of figure 2.1) we are dealing with a particular concept. In addition to particular concepts there are general or abstract concepts. For example, we can think about the spot in general—not as a particular, concrete spot (for example, one of those represented in the top row in figure 2.1) but about the spot as such. In the same way we can have an abstract concept of a straight line, a contour, a rectangle, a square, and so on.[2]

[2] According to the terminology accepted by many logicians, juxtaposing abstract concepts to concrete concepts is not at all the same as juxtaposing general concepts to particular ones. In a logic textbook (*Logika* [Logic], State Publishing House of Political Literature, Moscow, 1956) we read the following: "A concept by whose properties an object is conceived as such and as a given object is called concrete. A concept by whose properties what is conceived is not the given object as such but a certain property of the object or relationship among objects is called abstract."

This definition can hardly qualify as a masterpiece of clear thinking. Still, we may conclude from it that general concepts can also be considered abstract if they are formed not by listing particular objects included in them but rather by declaring a number of properties to be significant and abstracting from the other, insignificant properties. This is the only kind of general concepts we are going to consider and so we shall call them abstract concepts also. For example, an abstract triangle is any triangle regardless of its size, sides, or angles, or its position on the screening surface; therefore this is an abstract concept. The term "abstract" is used this way both in everyday life and in mathematics. At the same time, according to the logic textbook, "triangle," "square," and the like are concrete general concepts, but "triangularity" and "squareness," which are inherent in them, are abstract concepts. What is actually happening here is that a purely grammatical difference is being elevated to the rank of a logical difference, for, even from the point of view of an advocate of the latter variant of terminology, the possession of an abstract concept is equivalent to the possession of the corresponding general concept.

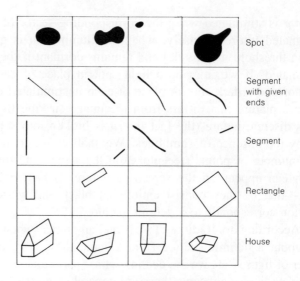

Figure 2.1. Pictures representing various concepts.

But what exactly does "possess an abstract concept" mean? how can we test whether someone possesses a given abstract concept—for example the concept of "spot"? There is plainly just one way: to offer the person being tested a series of pictures and ask him in each case whether or not it is a spot. If he correctly identifies each and every spot (and keep in mind that this is from the point of view of the test-maker) this means that he possesses the concept of spot. In other words, we must test his ability to *recognize* the affiliation of any picture offered with the set of pictures which we describe by the word "spot." Thus the abstract concept in the ordinary sense of the word (in any case, when we are talking about images perceived by the sense organs) coincides with the cybernetic concept we introduced—namely, that the concept is a set of situations. Endeavoring to make the term more concrete, we therefore call the task of recognition the task of pattern recognition, if we have in mind "generalized patterns" or the task of recognizing concepts, if we have in mind the recognition of particular instances of concepts.

In traditional logic the concrete concept of the "given picture" corresponds to a set consisting of one situation (picture). Relationships between sets have their direct analogs in relationships between concepts. If capital letters are used to signify concepts and small ones are used for the respective sets, the complement of set a, that is, the set of all conceivable situations not included in a, corresponds to the concept of "not A." The intersection of sets a, and b, that is, the set of situations which belong to both a and b, corresponds to the concept of "A and B simultaneously." For example, if A is the concept of "rectangle" and B is the concept of "rhombus," then "A and B simultaneously" is the concept of "square." The union of sets a and b, that is, the set of situations which belong to at least one of sets a and b, corresponds to the concept "either A, B, or A and B." If set a includes set b, that is, each element of b is included in a but the contrary is not true, then the concept B is a particular case of the concept A. In this case it is said that the concept A is more general (abstract) than the concept B, and the concept B is more concrete than A. For example, the square is a particular case of the rectangle. Finally, if sets a and b coincide then the concepts A and B are actually identical and distinguished, possibly, by nothing but the external form of their description, the method of recognition. Having adopted a cybernetic point of view, which is to say having equated the concept with a set of situations,, we should consider the correspondences enumerated above not as definitions of new terms but simply as an indication that there are several pairs of synonyms in our language.

■ DISCRIMINATORS AND CLASSIFIERS

WE SHALL CALL a nerve net that performs the task of recognition a *discriminator* (recognizer), and the state of the effector at its output will simply be called the state of the discriminator. Moving on from the concept of discriminator, we shall introduce the somewhat more general concept of *classifier*. The discriminator separates the set of all conceivable situations into two nonintersecting subsets: A and not-A. It is possible to imagine the division of a complete set of situations into an arbitrary number n of nonintersecting subsets. Such subsets are ordinarily called *classes*. Now let us picture a certain subsystem C

which has n possible states and is connected by a nerve net containing receptors in such a way that when a situation belongs to class i (concept i) the subsystem C goes into state i. We shall call such a subsystem and its nerve net a *classifier for a set of* n *concepts* (*classes*), and when speaking of the state of a classifier it will be understood that we mean the state of subsystem C (output subsystem). The discriminator is, obviously, a classifier with number of states $n = 2$.

In a system such as the nervous system, which is organized on the binary principle, the subsystem C with n sets will, of course, consist of a certain number of elementary subsystems with two states that can be considered the output subsystems (effectors) of the discriminators. The state of the classifier will, therefore, be described by indicating the states of a number of discriminators. These discriminators, however, can be closely interconnected by both the structure of the net and the function performed in the nervous system; in this case they should be considered in the aggregate as one classifier.

If no restrictions are placed on the number of states n the concept of the classifier really loses its meaning. In fact, every nerve net correlates one definite output state to each input state, and therefore a set of input states corresponds to each output state and these sets do not intersect. Thus, any cybernetic device with an input and an output can be formally viewed as a classifier. To give this concept a narrower meaning we shall consider that the number of output states of a classifier is many fewer than the number of input states so that the classifier truly "classifies" the input states (situations) according to a relatively small number of large classes.

■ HIERARCHIES OF CONCEPTS

FIGURE 2.2 shows a diagram of a classifier organized on the hierarchical principle. The *hierarchy* is, in general, that structure of a system made up of subsystems in which each subsystem is given a definite whole number, called its *level,* and the interaction of subsystems depends significantly on the difference in their levels according to some general principle. Ordinarily this principle is transmission of in-

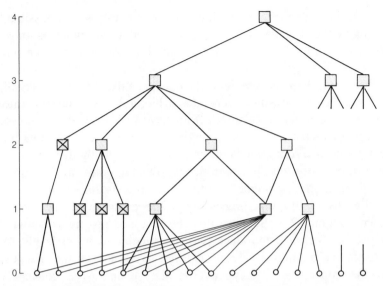

Figure 2.2. Hierarchy of classifiers.

formation in a definite direction, from top to bottom or bottom to top, from a given level to the next. In our case the receptors are called the zero level and the information is propagated from the bottom up. Each first-level subsystem is connected to a certain number of receptors and its state is determined by the states of the corresponding receptors. In the same way each second-level subsystem is connected with a number of first-level subsystems and so on. At the highest level (the fourth level in the diagram) there is one output subsystem, which gives the final answer regarding the affiliation of the situations with a particular class.

All subsystems at intermediate levels are also classifiers. The direct input for a classifier at level K is the states of the classifiers on level $K - 1$, the aggregate of which is the situation subject to classification on level K. In a hierarchical system containing more than one intermediate level, it is possible to single out hierarchical subsystems that bridge several levels. For example, it is possible to consider the states of all first-level classifiers linked to a third-level classifier as the input situation for the third-level classifier. Hierarchical systems can be added onto in breadth and height just as it is possible to put eight cubes together into a cube whose edges are twice as long as before. One can add more cubes to this construction to make other forms.

Because there is a system of concepts linked to each classifier the hierarchy of classifiers generates a hierarchy of concepts. Information is converted as it moves from level to level and is expressed in terms of increasingly "high-ranking" concepts. At the same time the amount of information being transmitted gradually decreases, because information that is insignificant from the point of view of the task given to the "supreme" (output) classifier is discarded.

Let us clarify this process with the example of the pictures shown in figure 2.1. Suppose that the assigned task is to recognize "houses." We shall introduce two intermediate concept levels. We shall put the aggregate of concepts of "segment" on the first level and the concept of "polygon" on the second. The concept of "house" comes on the third level.

By the concepts of "segment" we mean the aggregate of concepts of segments with terminal coordinates x_1, y_1 and x_2, y_2, where the numbers x_1, y_1, x_2 and y_2, can assume any values compatible with the organization of the screen and the system of coordinates. To be more concrete, suppose that the screen contains $1,000 \times 1,000$ light-sensitive points. Then the coordinates can be ten-digit binary numbers ($2^{10} = 1,024 > 1,000$), and a segment with given ends will require four such numbers, that is to say 40 binary orders, for its description. Therefore, there are 2^{40} such concepts in all. These are what the first-level classifiers must distinguish.

One should not think that a segment with given ends is a concrete concept—a set consisting of a single picture. When we classify this picture as a segment with given ends we are abstracting from the slight curvature of the line, from variations in its thickness, and the like (see figure 2.1). There are different ways to establish the criterion for determining which deviations from the norm should be considered insignificant. This does not interest us now.

Each first-level classifier should have at the output a subsystem of 40 binary orders on which the coordinates of the ends of the segment are "recorded." How many classifiers are needed? This depends on what kind of pictures are expected at the input of the system. Let us suppose that 400 segments are sufficient to describe any picture. This means that 400 classifiers are enough. We shall divide the entire screen into 400 squares of 50×50 points and link each square with a classifier which will fix a segment which is closest to it in some sense (the details of the division of labor among classifiers are insignificant). If there is no segment, let the classifier assume some conventional "meaningless" state, for example where all four coordinates are equal to 1,023.

If our system is offered a picture that shows a certain number of segments then the corresponding number of first-level classifiers will indicate the coordinates of the ends of the segments and the remaining classifiers will assume the state "no segment." This is a description of the situation in terms of the concepts of "segment." Let us compare the amount of information at the zero level and at the first level. At the zero level of our system $1,000 \times 1,000 = 10^6$ receptors receive 1 million bits of information. At the first level there are 400 classifiers, each of which contains 40 binary orders, that is, 40 bits of information; the total is 16,000 bits. During the transition to the first level the amount of information has decreased 62.5 times. The system has preserved only the information it considers "useful" and discarded the rest. The relativity of these concepts is seen from the fact that if the picture offered does not correspond to the hierarchy of concepts of the recognition system the system's reaction will be incorrect or simply meaningless. For example, if there are more than 400 segments in the picture not all of them will be fixed, and if a picture with a spot is offered the reaction to it will be the same as to any empty picture.

We divide the aggregate of concepts of "polygon," which occupies the second level of the hierarchy, into two smaller aggregates: isosceles triangles and parallelograms. We single out a special aggregate of rectangles from the parallelograms. Considering that assigning the angle and length requires the same number of bits (10) as for the coordinate, we find that 50 bits of information are needed to assign a definite isosceles triangle, 60 bits for a parallelogram, and 50 bits for a rectangle. The second-level classifiers should be designed accordingly. It is easy to see that all the information they need is available at the first level. The existence of a polygon is established where there are several segments that stand in definite relationships to one another. There is a further contraction of the information during the transition to the second level. Taking one third of the total of 400 segments for each type of polygon we obtain a system capable of fixing 44 triangles, 33 rectangles, and 33 parallelograms (simultaneously). Its information capacity is 5,830 bits, which is almost three times less than the capacity of the first level. On the other hand, when faced with an irregular triangle or quadrangle, the system is nonplussed!

It is easy to describe the concept of "house" in the language of second-level concepts. A house consists of four polygons—one rectangle, one isosceles triangle, and two parallelograms—which stand in definite relationships to one another. The base of the isosceles triangle coincides with one side of the rectangle, and so on.

To avoid misunderstanding it should be pointed out that the hierarchy of concepts we are discussing has a much more general meaning than the hierarchy of concepts by abstractness (generality) which is often simply called the "hierarchy of concepts." The pyramid of concepts used in classifying animals is an example of a hierarchy by generality. The separate individual animals (the "concrete" concepts) are set at the zero level. At the first level are the species, at the second the genera, then the orders families, classes, and phyla. At the peak of the pyramid is the concept of "animal." Such a pyramid is a particular case of the hierarchy of concepts in the general sense and is distinguished by the fact that each concept at level k is formed by joining a certain number of concepts at level $k - 1$. This is the case of very simply organized classifiers. In the general case classifiers can be organized any way one likes. The discriminators necessary to an animal are closer to a hierarchy based on complexity and subtlety of concepts, not generality.

■ HOW THE HIERARCHY EMERGES

LET US RETURN again to the evolution of the nervous system. Can a hierarchy of classifiers arise through evolution? It is apparent that it can, but on one condition: if the creation of each new level of the hierarchy and its subsequent expansion are useful to the animal in the struggle for existence. As animals with highly organized nervous systems do exist, we may conclude that such an expansion is useful. Moreover, studies of primitive animals show that the systems of concepts their nervous systems are capable of recognizing are also very primitive. Consequently, we see for ourselves the usefulness of the lowest level of the hierarchy of classifiers.

Let us sketch the path of development of the nervous system. In the initial stages we find that the animal has just a few receptors. The number of possible methods of interconnecting them (combinations) is relatively small and permits direct selection. The advantageous combination is found by the trial and error method. That an advantageous combination can exist even for a very small number of neurons can easily be seen in the following example. Suppose that there are just two light-sensitive receptors. If they are set on different sides of the body the information they yield (difference in illuminations) is sufficient for the animal to be able to move toward or away from the light. When an advantageous combination has been found and realized by means, we shall assume, of one intermediate neuron (such neurons are called *associative*), the entire group as a whole may be reproduced. In this way there arises a system of associative neurons which, for example, register differences between the illumination of receptors and sum these differences (see figure 2.3a). Any part of a system of connected neurons may be reproduced, for example one or several receptors. In this way there arises a system of connections of the type shown in figure 2.3b. The diagrams of both types taken together form the first level of a hierarchy, based on the concepts of the sum and difference of illuminations. Because it is very important that animal movement be able to adjust to changes in illumination at a given point, we may assume that neurons capable of being triggered by changes in illumination must have appeared in the very early stages of evolution. They could have been either receptors or associative neurons connected to one or several receptors. In general, first-level classifiers can be described as registers of the sum and differences of the stimuli of receptors in space and time.

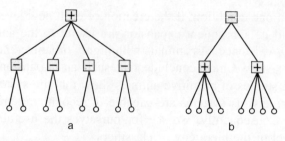

Figure 2.3. Simplest types of connections among receptors.

Having proven their usefulness for the animal, first-level classifiers become an established part of its capabilities in the struggle for existence. Then the next trial and error series begins: a small number of first-level classifiers (to be more precise, their output subsystems) are interconnected into one second-level trial classifier until a useful combination is obtained. Then the reproduction of this combination is useful. It may be assumed that on the second level of the hierarchy (pertaining to the organs of sight) there appear such concepts as the boundary between light and shadow, the spot, the average illumination of a spot, and movement of the boundary between light and shadow. The successive levels of the hierarchy will arise in the same way.

The scheme we have outlined leads one to think that any complex system which has arisen by the method of trial and error in the process of evolution should have a hierarchical organization. In fact, nature—unable to sort through all conceivable combinations of a large number of elements—selects combinations from a few elements. When it finds a useful combination, nature reproduces it and uses it (the whole of it) as an element to be tentatively connected with a small number of other similar elements. This is how the hierarchy arises. This concept plays an enormous role in cybernetics. In fact, any complex system, whether it has arisen naturally or been created by human beings, can be considered organized only if it is based on some kind of hierarchy or interweaving of several hierarchies. At least we do not yet know any organized systems that are arranged differently.

■ SOME COMMENTS ON REAL HIERARCHIES

THUS FAR our conclusions have been purely speculative. How do they stand up against the actual structure of the nervous systems of animals and what can be said about the concepts of intermediate levels of a hierarchy which has actually emerged in the process of evolution?

When comparing our schematic picture with reality the following must be considered. The division of a system of concepts into levels is not so unconditional as we have silently assumed. There may be cases where concepts on level K are used directly on level $K + 2$, bypassing level $K + 1$. In figure 2.2 we fitted such a possibility into the overall diagram by introducing classifiers which are connected to just one classifier of the preceding level and repeat its state; they are shown by the squares containing the x's. In reality, of course, there are no such squares, which complicates the task of breaking the system up into levels. To continue, the hierarchy of classifiers shown in figure 2.2. has a clearly marked pyramidal character; at higher levels there are fewer classifiers and at the top level there is just one. Such a situation occurs when a system is extremely "purposeful," that is, when it serves some very narrow goal, some precisely determined method of classifying situations. In the example we have cited this was recognition of "houses." And we saw that for such a system even irregular triangles and quadrangles proved to be "meaningless"; they are not included in the hierarchy of concepts. To be more universal a system must resemble not one pyramid but many pyramids whose apexes are arranged at approximately the same level and form a set of concepts (more precisely, a set of systems of concepts) in whose terms control of the animal's actions takes place and which are ordinarily discovered during investigation of the animal's behavior. These concepts are said to form the basis of a definite "image" of the external world which takes shape in the mind of the animal (or person). The state of the classifiers at this level is direct information for the executive part of the nerve net (that is, in the end, for the effectors). Each of these classifiers relies on a definite hierarchy of classifiers, a pyramid in which information moves as described above. But the pyramids may overlap in their middle parts (and they are known to overlap in the lower part, the receptors). Theoretically the total number of pyramid apexes may be as large as one likes, and specifically it may be much greater than the total number of receptors. This is the case in which the very same information delivered by the receptors is represented by a set of pyramids in a set of different forms figured for all cases in life.

Let us note one other circumstance that should be taken into account in the search for hierarchy in a real nerve net. If we see a neuron connected by synapses with a hundred receptors, this by itself does not mean that the neuron fixes some simple first-level concept such as the total number of stimulated receptors. The logical function that relates the state of the neuron to the states of the receptors may be very complex and have its own hierarchical structure.

■ THE WORLD THROUGH THE EYES OF A FROG

FOUR SCIENTISTS from the Massachusetts Institute of Technology (J. Lettvin et al.) have written an article entitled "What the Frog's Eye Tells the Frog's Brain" which is extremely interesting for an investigation of the hierarchy of classifiers and concepts in relation to visual perception in animals.[3] The authors selected the frog as their test animal because its visual apparatus is relatively simple, and therefore convenient for study. Above all, the retina of the frog eye is homogeneous; unlike the human eye it does not have an area of increased sensitivity to which the most important part of the image must be projected. Therefore, the glance of the frog is immobile; it does not follow a moving object with its eyes the way we do. On the other hand, if a frog sitting on a water lily rocks with the motion of the plant, its eyes make the same movements, thus compensating for the rocking, so that the image of the external world on the retina remains immobile. Information is passed from the retina along the visual

---

[3] See the Russian translation in the collection of articles entitled *Elektronika i kibernetika v biologii i meditsine* (Electronics and Cybernetics in Biology and Medicine), Foreign Literature Publishing House, Moscow, 1963. [Original Lettvin et al., *Proc. IRE*, 47, 1940–1951 (1959, #11)].

nerve to the so-called thalamus opticus of the brain. In this respect the frog is also simpler than the human being; the human being has two channels for transmitting information from the retina to the brain.

Vision plays a large part in the life of the frog, enabling it to hunt and to protect itself from enemies. Study of frog behavior shows that the frog distinguishes its prey from its enemies by size and state of movement. Movement plays the decisive part here. Having spotted a small moving object (the size of an insect or worm) the frog will leap and capture it. The frog can be fooled by a small inedible object wiggled on a thread, but it will not pay the slightest attention to an immobile worm or insect and can starve to death in the middle of an abundance of such food if it is not mobile. The frog considers large moving objects to be enemies and flees from them.

The retina of the frog's eye, like that of other vertebrates, has three layers of nerve cells. The outermost layer is formed by light-sensitive receptors, the rods and cones. Under it is the layer of associative neurons of several types. Some of them (the *biopolar cells*) yield primarily vertical axons along which stimulation is transmitted to deeper layers. The others (the *horizontal* or *amacrine* cells) connect neurons that are located on one level. The third, deepest layer is formed of the *ganglion* cells. Their dendrites receive information from the second-layer cells and the axons are long fibers that are interwoven to form the visual nerve, which connects the retina with the brain. These axons branch out, entering the thalamus opticus, and transmit information to the dendrites of the cerebral neurons.

The eye of a frog has about 1 million receptors, about 3 million associative second-level neurons, and about 500,000 ganglion cells. Such a retinal structure gives reason to assume that analysis of the image begins in the eye of the animal and that the image is transmitted along the visual nerve in terms of some intermediate concepts. It is as if the retina were a part of the brain moved to the periphery. This assumption is reinforced by the fact that the arrangement of the axons on the surface of the thalamus opticus coincides with the arrangement of the respective ganglion cells at the output of the retina—even though the fibers are interwoven a number of times along the course of the visual nerve and change their position in a cross-section of the nerve! Finally, the findings of embryology on development of the retina lead to the same conclusion.

In the experiments we are describing a thin platinum electrode was applied to the visual nerve of a frog, making it possible to record stimulation of separate ganglion cells. The frog was placed in the center of an aluminum hemisphere, which was dull gray on the inside. Various dark objects such as rectangles, discs, and the like, were placed on the inside surface of the hemisphere; they were held in place by magnets set on the outside.

The results of the experiments can be summarized as follows. Each ganglion cell has a definite *receptive field,* that is, a segment of the retina (set of receptors) from which it collects information. The state of receptors outside the receptive field has no effect on the state of the ganglion cell. The dimensions of receptive fields for cells of different types, if they are measured by the angle dimensions of the corresponding visible areas, vary from 2 degrees to 15 degrees in diameter.

The ganglion cells are divided into four types depending on what process they record in their receptive field.

1. Detectors of long-lasting contrast. These cells do not react to the switching on and off of general illumination, but if the edge of an object which is darker or lighter than the background appears in the receptive field the cell immediately begins to generate impulses.

2. Detectors of convex edges. These cells are stimulated if a small (not more than three degrees) convex object appears in the receptive field. Maximum stimulation (frequency of impulses) is reached when the diameter of the object is approximately half of the diameter of the receptive field. The cell does not react to the straight edge of an object.

3. Detectors of moving edges. Their receptive fields are about 12 degrees in width. The cell reacts to any distinguishable edge of an object which is darker or lighter than the background, but only if it is moving. If a smoothly moving object five degrees in width passes over the field there are two reactions, to the front and rear edges.

4. Detectors of darkening of the field. They send out a series of

impulses if the total illumination of the receptive field is suddenly decreased.

The arrangement of the ends of the visual fibers in the thalamus opticus is extremely interesting. We have already said that on a plane this arrangement coincides with the arrangement of the corresponding ganglion cells in the retina. In addition, it turns out that the ends of each type of fiber are set at a definite depth in the thalamus opticus, so that the frog brain has four layers of neurons that receive visual information. Each layer receives a copy of the retinal image—but in a certain aspect that corresponds to one of the four types of ganglion cells. These layers are the transmitters of information for the higher parts of the brain.

Experiments such as those we have described are quite complex and disputes sometimes arise concerning their interpretation. The details of the described system may change or receive a different interpretation. Nonetheless, the general nature of the system of first-level concepts has evidently been quite firmly established. We see a transition from point description to local description which takes account of the continuous structure of the image. The ganglion cells act as recognizers of such primary concepts as edge, convexness, and movement in relation to a definite area of the visible world.

■ FRAGMENTS OF A SYSTEM OF CONCEPTS

THE LOWEST-LEVEL concepts related to visual perception for a human being probably differ little from the concepts of a frog. In any case, the structure of the retina in mammals and in human beings is the same as in amphibians.

The phenomenon of distortion of perception of an image stabilized on the retina gives some idea of the concepts of the subsequent levels of the hierarchy. This is a very interesting phenomenon. When a person looks at an immobile object, "fixes" it with his eyes, the eyeballs do not remain absolutely immobile; they make small involuntary movements. As a result the image of the object on the retina is constantly in motion, slowly drifting and jumping back to the point of

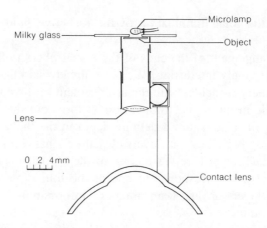

Figure 2.4. Device for stabilizing an image on the retina.

maximum sensitivity. The image "marks time" in the vicinity of this point.

An image which is stabilized, not in continuous motion, can be created on the retina. To achieve this, the object must be rigidly connected to the eyeball and move along with it (see figure 2.4). A contact lens with a small rod secured to it is placed on the eye. The rod holds a miniature optical projector [4] into which slides a few millimeters in size can be inserted. The test subject sees the image as remote to the point of infinity. The projector moves with the eye so the image on the retina is immobile.

When the test subject is shown a stabilized image, for the first few seconds he perceives it as he would during normal vision, but then distortions begin. First the image disappears and is replaced by a gray or black background, then it reappears in parts or whole.

That the stabilized image is perceived incorrectly is very remarkable in itself. Logically, there is no necessity for the image of an immobile object to move about the retina. Such movement produces no increase in the amount of information, and it becomes more difficult to process it. As a matter of fact, when similar problems arise in the

[4] See R. Pritchard, "Images on the Retina and Visual Perception," in the collection of articles *Problemy bioniki* (Problems of Bionics), Mir Publishing House, 1965. [Original in English "Stabilized Images on the Retina," *Scientific American* 204 no. 41 (June 1961): 72–78].

area of engineering—for example when an image is transmitted by television or data are fed from a screen to a computer—special efforts are made to stabilize the image. But the human eye has not merely adapted to a jerking image; it simply refuses to receive an immobile one. This is evidence that the concepts related to movement, probably like those which we observed in the frog, are deeply rooted somewhere in the lower stages of the hierarchy, and if the corresponding classifiers are removed from the game correct information processing is disrupted. From the point of view of the designer of a complex device such as the eye (plus the data processing system) such an arrangement is strange. The designer would certainly fill all the lower stages with static concepts and the description of object movement would be given in terms of the concepts of a higher level. But the hierarchy of visual concepts arose in the process of evolution. For our remote frog-like ancestors the concepts related to movement were extremely important and they had no time to wait for the development of complex static concepts. Therefore, primitive dynamic concepts arose in the very earliest stages of the development of the nervous system, and because nature uses the given units to carry out subsequent stages of building, these concepts became firmly established at the base of the hierarchy of concepts. For this reason, the human eyeball must constantly make brownian movements.

Even more interesting is the way the image breaks up into parts (*fragmentation*). Simple figures, such as a lone segment, disappear and come back in toto. More complex figures sometimes disappear in toto and sometimes break into parts which disappear and reappear independent of one another (see figure 2.5). Fragmentation does not occur chaotically and it is not independent of the type of image, as is the case when a drawing on a chalkboard is erased with a rag; rather the fragmentation corresponds to the "true" structure of the image. We have put the word "true" in quotation marks because fragmentation actually occurs in accordance with the structure of image perception by the eye–brain system. We do not know exactly what the mechanics of the distortion of perception in stabilization are; we know only that stabilization disables some component of the perception system. But from this too we can draw certain conclusions.

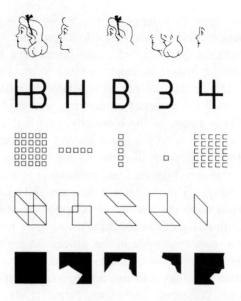

Figure 2.5. Fragmentation of a stabilized image.

Imagine that several important design elements have suddenly disappeared from an architectural structure. The building will fall down, but probably the pieces would be of very different sizes. In one place you may see individual bricks and pieces of glass, while in another a part of the wall and roof may remain, and in still another place a whole corner of the building may be intact. Perception of the stabilized image is approximately that kind of sight. It makes it possible to picture the nature of the concepts of a higher level (or higher levels) but not to evaluate their mutual relations and dependences. It should be noted that in the human being the personal experience of life, the *learning* (to speak in cybernetic language), plays a large part in shaping higher-level concepts. (This will be the next stage in evolution of the nervous system, so we are getting somewhat ahead of things here. For an investigation of the hierarchy of concepts, however, it is not very important whether the hierarchy were inherited or acquired through one's own labor.)

Let us cite a few excerpts from the work mentioned above (footnote 4).

The figure of the human profile invariably fades and regenerates in meaningful units. The front of the face, the top of the head, the eye and ear come and go as recognizable entities, separately and in various combinations. In contrast, on first presentation a meaningless pattern of curlicues is described as extremely "active"; the individual elements fade and regenerate rapidly, and the subject sees almost every configuration that can be derived from the original figure. After prolonged viewing, however, certain combinations of curlicues become dominant and these then disappear and reappear as units. The new reformed groupings persist for longer periods. . . .

Linear organization is emphasized by the fading of this target composed of rows of squares. The figure usually fades to leave one whole row visible: horizontal, diagonal, or vertical. In some cases a three-dimensional "waffle" effect is also noted. . . .

A random collection of dots will fade to leave only those dots which lie approximately in a line. . . . Lines act independently in stabilized vision, with breakage in the fading figure always at an intersection of lines. Adjacent or parallel lines may operate as units. . . . In the case of figures drawn in solid tones as distinguished from those drawn in outline . . . the corner now replaces the line as the unit of independent action. A solid square will fade from its center, and the fading will obliterate first one and then another corner, leaving the remaining corners sharply outlined and isolated in space. Regeneration correspondingly begins with the reappearance of first one and then another corner, yielding a complete or partial figure with the corners again sharply outlined.

■ THE GOAL AND REGULATION

WE HAVE DESCRIBED the first half of the action of a complex reflex, which consists of analyzing the situation by means of a hierarchy of classifiers. There are cases where the second half, the executive half, of the reflex is extremely simple and involves the stimulation of some local group of effectors—for example the effectors that activate a certain gland. These were precisely the conditions in which I. P. Pavlov set up most of his experiments, experiments which played an important part in the study of higher nerve activity in animals and led to his widely known theory of unconditioned and conditioned reflexes. Elementary observations of animal behavior under natural conditions show, however, that this behavior cannot be reduced to a set of reflexes that are related only to the state of the environment. Every action of any complexity whatsoever consists of a sequence of simpler actions joined by a common goal. It often happens that individual components in this aggregate of actions are not simply useless but actually harmful to the animal if they are not accompanied by the other components. For example, it is necessary to fall back on the haunches before jumping and in order to grasp prey better it must be let go for an instant. The two phases of action, preliminary and executive, which we see in these examples cannot be the result of independent reflexes because the first action is senseless by itself and therefore could not have developed.

When describing behavior the concepts of *goal* and *regulation* must be added to the concept of the reflex. A diagram of regulation is shown in figure 2.6. An action which the system is undertaking depends not only on the situation itself but also on the *goal,* that is, on the situation that the system is trying to achieve. The action of the system is determined by comparing the situation and the goal: the action is directed toward eliminating the discrepancy between the situation and the goal. The situation determines the action through the comparison block. The action exerts a reverse influence on the situation through change in the environment. This feedback loop is a typical feature of the regulation diagram and distinguishes it from the reflex diagram where the situation simply causes the action.

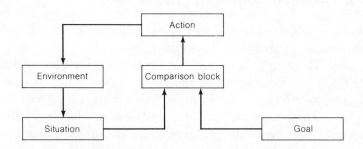

Figure 2.6. Diagram of regulation.

■ HOW REGULATION EMERGES

HOW COULD A SYSTEM organized according to the regulation diagram occur in the process of evolution? We have seen that the appearance of hierarchically organized classifiers can be explained as a result of the combined action of two basic evolutionary factors: replication of biological structures and finding useful interconnections by the trial and error method. Wouldn't the action of these factors cause the appearance of the regulation diagram?

Being unable to rely on data concerning the actual evolutionary process that millions of years ago gave rise to a complex nervous system, we are forced to content ourselves with a purely hypothetical combinative structure which demonstrates the theoretical possibility of the occurrence of the regulation diagram. We shall make a systematic investigation of all possibilities to which replication and selection lead. It is natural to assume that in the process of replication relations are preserved within the subsystem being replicated, as are the subsystem's relations with those parts not replicated. We further assume that owing to their close proximity there is a relationship among newly evolved subsystems, which we shall depict in our diagrams with a dotted line. This relationship may either be reinforced or disappear. We shall begin with the simplest case—where we see just one nerve cell that is receptor and effector at the same time (figure 2.7a). Here there is only one possibility of replication, and it leads to the appearance of two cells (figure 2.7b). If one of them is closest to the surface and the other closer to the muscle cells, a division of labor between them is useful. This is how the receptor-effector diagram emerges (figure 2.7c).

Now two avenues of replication are possible. Replication of the receptor yields the pattern shown in figure 2.7d; after the disappear-

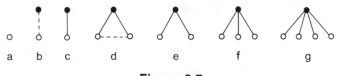

a   b   c   d   e   f   g

Figure 2.7

ance of the dotted-line relationship, this becomes figure 2.7e. A similar process generates the patterns in figures 2.7f, g, and so on. In this way the zero level of the hierarchy (receptors) expands.

The second avenue is replication of effectors (see figure 2.8). In figure 2.8b, the stimulation of one receptor should be transmitted along two channels to two effectors. But we know that the electrical resistance of the synapses drops sharply after the first time a current passes along them. Therefore, if the stimulation is sent along one channel this communications channel will be reinforced while the other will be bypassed and may "dry up" (figure 2.8c). Then the stimulation may make a way across the dotted-line relationship (figure 2.8d), which marks the birth of the first level of the hierarchy of classifiers.

Figure 2.9 shows possible variations of the development of the three-neuron diagram shown in figure 2.7d. The diagrams correspond to replication of different subsystems of the initial system. The subsystem which is replicated has been circled. Figures 2.9a–c explain the expansion of the zero level, while figures 2.9d–f show the expansion of the first two levels of the hierarchy of classifiers. In the remainder we see patterns that occur where one first-level classifier is replicated without a receptor connected to it. The transition from figure 2.9h to 2.9i is explained by that "drying up" of the bypass channel we described above. Figure 2.9j, the final development, differs substantially from all the other figures that represent hierarchies of classifiers. In this figure, one of the classifiers is "hanging in the air"; it does not receive information from the external world. Can such a diagram be useful to an animal? It certainly can, for this is the regulation diagram!

As an example we can suggest the following embodiment of fig-

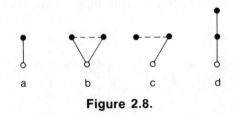

a   b   c   d

Figure 2.8.

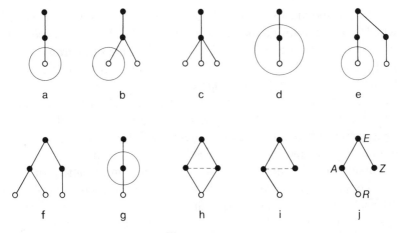

Figure 2.9.

ure 2.9j. Let us consider a certain hypothetical animal which lives in the sea. Suppose R is a receptor which perceives the temperature of the environment. Classifier A also records the temperature of the water by change in the frequency of stimulation impulses. Suppose that greater or less stimulation of effector E causes expansion or contraction of the animal's shell, which results in a change in its volume; the animal either rises toward the surface of the sea or descends deeper. And suppose that there is some definite temperature, perhaps 16° C (61° F) which is most suitable for our animal. The neuron Z (the goal fixer) should maintain a certain frequency of impulses equal to the frequency of neuron A at a temperature of 16°. Effector E should register the difference of stimulation of neurons A and Z and, in conformity with it, raise the animal toward the surface where the water is warmer or immerse it to deeper, cooler water layers. Such an adaptation would be extremely helpful to our imaginary animal.

■ REPRESENTATIONS

REPLICATION of the various subsystems of the nerve net can give rise to many different groups of classifiers which "hang in the air." Among them may be copies of whole steps of the hierarchy whose states correspond exactly to the states of those "informed" classifiers which receive information from the receptors. They correspond but they do not coincide. We saw this in the example of neurons A and Z in Figure 2.9j. In complex systems the uninformed copies of informed classifiers may store a large amount of information. We shall call the states of these copies *representations,* fully aware that in this way we are giving a definite cybernetic interpretation to this psychological concept. It is obvious that there is a close relationship between representations and situations, which are really nothing but the states of analogous classifiers, but ones receiving information from the receptors. The goal is a particular case of the representation, or more precisely, it is that case where the comparison between a constant representation and a changing situation is used to work out an action that brings them closer to one another. The hypothetical animal described above loves a temperature of 16° and the "lucid image" of this wonderful situation, which is a certain frequency of impulses of neuron A, lives in its memory in the form of precisely that frequency of pulses of neuron Z.

This is a very primitive representation. The more highly organized the "informed" part of the nervous system is, the more complex its duplicates will be (we shall call them *representation fixers*), and the more varied the representations will be. Because classifiers can belong to different levels of the hierarchy and the situation can be expressed in different systems of concepts, representations can also differ by their "concept language" because they can be the states of fixers of different levels. Furthermore, the degree of stability of the states of the representation fixers can also vary greatly. Therefore, representations differ substantially in their concreteness and stability. They may be exact and concrete, almost perceptible to the sensors. The extreme case of this is the hallucination, which is perceived subjectively as reality and to which the organism responds in the same way as it would to the corresponding situation. On the other hand, representations may be very approximate, as a result of both their instability and their abstraction. The latter case is often encountered in artistic and scientific creative work where a representation acts as the goal of activity. The human being is dimly aware of what he needs and tries to embody it in solid, object form. For a long time nothing comes of it because his representations do not have the necessary concreteness. But then, at one fine moment (and this is really a *fine* moment!) he suddenly achieves his goal and realizes clearly that he has done precisely what he wanted.

■ MEMORY

IN PRINCIPLE, as many representation fixers as desired can be obtained by replication. But a question arises here: how many does an animal need? How many copies of "informed" classifiers are needed? One? Two? Ten? It follows from general considerations that many copies are needed. After all, representation fixers serve to organize experience and behavior in time. The goal fixer stores the situation which, according to the idea, should be realized in the future. Other fixers can store situations which have actually occurred in the past. The temporal organization of experience is essential to an animal which is striving to adapt to the environment in which it lives, for this environment reveals certain rules, that is, correlations between past, present, and future situations. We may predict that after a certain initial increase in the number of receptors the further refinement of the nervous system will require the creation of representation fixers, and a large number of them. There is no reason to continue to increase the number of receptors and classifiers and thus improve the "instantaneous snapshots" of the environment if the system is not able to detect correlations among them. But the detection of correlations among the "instantaneous snapshots" requires that they be stored somewhere. This is how representation fixers, which in other words are memory, arise. The storage of the goal in the process of regulation is the simplest case of the use of memory.

■ THE HIERARCHY OF GOALS AND PLANS

IN THE REGULATION DIAGRAM in figure 2.5 the goal is shown as something unified. But we know very well that many goals are complex, and while working toward them a system sets intermediate goals. We have already cited the examples of two-phase movement: to jump onto a chair, a cat first settles back on its haunches and then springs up. In more complex cases the goals form a hierarchy consisting of numerous levels. Let us suppose that you set the goal of traveling from home to work. This is your highest goal at the particular moment. We shall assign it the index (level number) 0. To travel to work you must leave the building, walk to the bus stop, ride to the necessary stop, and so on. These are goals with an index of -1. To leave the building you must leave the apartment, take the elevator down, and go out the entrance. These are goals with an index of -2. To take the elevator down you must open the door, enter the elevator, and so on; this is index -3. To open the elevator door you must reach your hand out to the handle, grasp it, and pull it toward you; this is index -4. These goals may perhaps be considered elementary.

The goal and a statement of how it is to be achieved—that is, a breakdown into subordinate goals—is called a *plan* of action. Our example is in fact a description of a *plan* for traveling to work. The goal itself, which in this case is the representation "me—at my work place," does not have any hierarchical structure. The primary logical unit that forms the hierarchy is the plan, but the goals form a hierarchy only to the extent that they are elements of the plan.

In their book *Plans and the Structure of Behavior* American psychologists G. Miller, E. Galanter, and K. Pribram take the concept of the plan as the basis for describing the behavior of humans and animals. They show that such an approach is both sound and useful. Unlike the classical reflex arc (without feedback) the logical unit of behavior description used by the authors contains a feedback loop (see figure 2.10). They call this unit the Test-Operate-Test-Exit diagram (T-O-T-E—based on the first letters of the English words "test," "operate," "test," "exit.") The test here means a test of correspondence between the situation and the goal. If there is no correspondence an operation is performed, but if there is correspondence the plan is considered performed and the system goes to "exit." As an example, figure 2.11 shows a plan for driving a nail into a board; the plan is represented in the form of a T-O-T-E unit. The T-O-T-E diagram in figure 2.10 shows the same phenomenon of regulation that

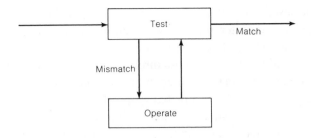

Figure 2.10. T-O-T-E. Test-operate-test-exit unit.

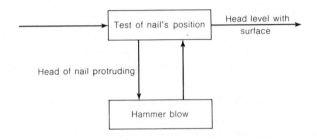

Figure 2.11. Driving a nail.

was depicted in figure 2.6. The difference is in the method of depiction. The diagram in figure 2.6 is structural while in figure 2.10 it is functional. We shall explain these concepts, and at the same time we shall define the concept of control more precisely.

■ STRUCTURAL AND FUNCTIONAL DIAGRAMS

A STRUCTURAL DIAGRAM of a cybernetic system shows the subsystems which make up the particular system and often also indicates the directions of information flows among the subsystems. Then the structural diagram becomes a graph. In mathematics the term *graph* is used for a system of points (the *vertices* of the graph), some of which are connected by lines (*arcs*). The graph is *oriented* if a definite direction is indicated on each arc. A structural diagram with an indication of information flows is a directed graph whose vertices depict the subsystems while the arcs are the information flows.

This description of a cybernetic system is not the only possible one. Often we are interested not so much in the structure of a system as in its functioning. Even more often we are simply unable to say anything sensible about the structure, but there are some things we can say about the function. In such cases a *functional diagram* may be constructed. It is also a directed graph, but in it the vertices represent different sets of states of the system and the arcs are possible transitions between states. An arc connects to vertices in the direction from the first to the second in the case where there is a possibility of transition from at least one state relating to the first vertex into another state relating to the second vertex. We shall also call the sets of states *generalized states*. Therefore, the arc in a diagram shows the possibility of a transition from one generalized state to another. Whereas a structural diagram primarily reflects the spatial aspect, the functional diagram stresses the temporal aspect. Formally, according to the definition given above, the functional diagram does not reflect the spatial aspect (division of the system into subsystems) at all. As a rule, however, the division into subsystems is reflected in the method of defining generalized states, that is, the division of the set of all states of the system into subsets which are "assigned" to different vertexes of the graph. Let us review this using the example of the system whose structural diagram is shown in figure 2.12. This is a *control diagram*.

One of the subsystems, which is called the *control device,* receives information from "working" subsystems A_1, A_2, A_3, . . . , processes it, and sends *orders* (control information) to subsystems A_1, A_2, A_3, . . . , as a result of which these subsystems change their

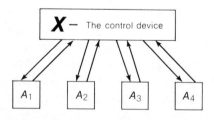

Figure 2.12. Structural diagram of control.

50   HIERARCHICAL STRUCTURES

state. It must be noted that, strictly speaking, any information received by the system changes its state. Information is called control information when it changes certain distinct parameters of the system which are identified as "primary," "external," "observed," and the like.

Often the control unit is small in terms of information-capacity and serves only to switch information flows, while the real processing of data and development of orders is done by one of the subsystems, or according to information stored in it. Then it is said that control is passed to this subsystem. That is how it is done, specifically, in a computer where subsystems A_1, A_2, A_3, . . . are the cells of operational memory. Some of the cells contain "passive" information (for example numbers), while others contain orders (instructions). When control is in the cell which contains an instruction the control unit performs this instruction. Then it passes control to the next cell, and so on.

The functional diagram for systems with transfer of control is constructed as follows. To each vertex of the graph is juxtaposed one of the subsystems A_i and the set of all states of the system when control is in the particular subsystem. Then the arcs (arrows) signify the transfer of control from one subsystem to another (see figure 2.13). Even where each successive state is fully determined by the preceding one there may be branching on such a diagram because each vertex corresponds to a vast number of states and the tranfer of control

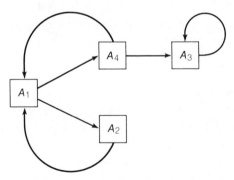

Figure 2.13. Functional diagram of transfer of control.

can depend on the state of the control unit or the subsystem in which control is located. Functional diagrams are often drawn in generalized form, omitting certain inconsequential details and steps. It may then turn out that the path by which control branches depends on the state of several different subsystems. The condition on which this switch is made is ordinarily written alongside the arrow. The diagram shown in figure 2.10 can be understood in precisely this sense. Then it will be assumed that the system has two subsystems, a test block and an operation-execution block, and control passes from one to the other in conformity with the arrows. The system can also have other subsystems (in this case the environment), but they never receive control and therefore are not shown in the diagram (to be more precise, those moments when the environment changes the state of the system or changes its own state when acted upon by the system are included in the process of action of one of the blocks).

We can move even further from the structural diagram. Switching control to a certain subsystem means activating it, but there can be cases where we do not know exactly which subsystem is responsible for a particular observed action. Then we shall equate the vertices of the graph with the actions as such and the arcs will signify the transition from one action to another. The concept of "action as such," if strictly defined, must be equated with the concept of "generalized state" ("set of states") and this returns us to the first, most abstract definition of the functional diagram. In fact, when we say that a dog "runs," "barks," or "wags his tail," a set of concrete states of the dog fits each of these definitions. Of course one is struck by a discrepancy. "State" is something static, but "action" is plainly something dynamic, closer to a change of state than a state itself. If a photograph shows a dog's tail not leaving the plane of symmetry, we still do not know whether the dog is wagging it or holding it still. We overcome such contradictions by noting that the concept of state includes not only quantities of the type "position," but also quantities such as "velocity," "acceleration," and the like. Specifically, a description of the state of the dog includes an indication of the tension of its tail muscles and the stimulation of all neurons which regulate the state of the muscles.

■ THE TRANSITION TO PHENOMENOLOGICAL DESCRIPTIONS

THEREFORE, in the functional diagram an action is, formally speaking, a set of states. But to say that a particular action is some set is to say virtually nothing. This set must be defined. And if we do not know the structure of the system and its method of functioning it is practically impossible to do this with precision. We must be content with an incomplete, phenomenological definition based on externally manifested consequences of internal states. It is this kind of functional diagram, with more or less exactly defined actions at the vertices of the graph, that is used to describe the behavior of complex systems whose organization is unknown—such as humans and animals. The diagrams in figures 2.10 and 2.11 are, of course, such diagrams. The phenomenological approach to brain activity can be carried out by two sciences: psychology and behavioristics (the study of behavior). The former is based on subjective observations and the latter on objective ones. They are closely connected and are often combined under the general name of psychology.

Because the operational component of the T-O-T-E unit may be

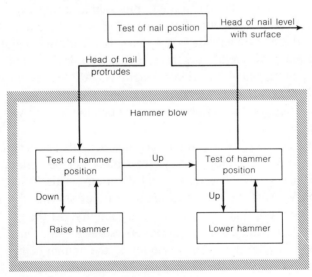

Figure 2.14. Hierarchical plan for driving a nail.

composite, requiring the performance of several subordinate plans, T-O-T-E units can have hierarchical structure. Miller, Galatier, and Pribram give the following example. If a hammer striking a nail is represented as a two-phase action consisting of raising and lowering the hammer, then the functional diagram in figure 2.11 which depicts a plan for driving a nail, becomes the diagram shown in figure 2.14. In its turn, this diagram can become an element of the operational component of a T-O-T-E diagram on a higher level.

We have seen that the elementary structural diagram of figure 2.6 corresponds to the elementary functional diagram in figure 2.9. When plans make up the hierarchy, what happens to the structural diagram? Or, reversing the statement to be more precise, what structural diagrams can ensure execution of a hierarchically constructed plan?

Different variants of such diagrams may be suggested. For example, it can be imagined that there is always one comparison block and that the same subsystem which stores the goal is always used, but the state of this subsystem (that is, the goal) changes under the influence of other parts of the system, ensuring an alternation of goals that follows the plan. By contrast, it may be imagined that the comparison block–goal pair is reproduced many times and, during execution of a hierarchical plan, control passes from one pair to the other. A combination of these two methods may be proposed and, in general, we can think up many differently organized cybernetic devices that carry out the same hierarchical functional diagrams. All that is clear is that they will have a hierarchical structure and that devices of this type can arise through evolution by the replication of subsystems and selection of useful variants.

But what kind of structural diagrams actually appear in the process of evolution? Unfortunately, we cannot yet be certain. That is why we had to switch to functional diagrams. This is just the first limitation we shall be forced to impose on our striving for a precise cybernetic description of higher nervous activity. At the present time we know very little about the cybernetic structure and functioning of the brains of higher animals, especially of the human being. Properly speaking, we know virtually nothing. We have only certain facts and assumptions. In our further analysis, therefore, we shall have to rely on phenomenology, the findings of behavioristics and psychology, where things are somewhat better. As for the cybernetic aspect, we shall move to the level of extremely general concepts, where we shall find certain rules so general that they explain the stages of development of both the nervous system and human culture, in particular science. The relatively concrete cybernetic analysis of the first stages of evolution of the nervous system, which is possible thanks to the present state of knowledge, will serve as a running start for the subsequent, more abstract analysis. Of course, our real goal is precisely this abstract analysis, but it would be more satisfying if, knowing more about the cybernetics of the brain, we were able to make the transition from the concrete to the abstract in a more smooth and well-substantiated manner.

■ DEFINITION OF THE COMPLEX REFLEX

SUMMARIZING our description of the fourth stage in the development of the nervous system we can define the complex reflex as that process where stimulation of receptors caused by interaction with the environment is passed along the nerve net and is converted by it, thus activating a definite plan of action that immediately begins to be executed. In this diagram of behavior all feedbacks between the organism and the environment are realized in the process of regulation of actions by the plan, while overall interaction between the environment and the organism is described by the classical stimulus–response formula. Only now the response means activation of a particular plan.

CHAPTER THREE
On the Path toward the Human Being

■ THE METASYSTEM TRANSITION

SUBSEQUENT STAGES in the development of the nervous system will be described as stated above, on a more phenomenological level. For this we must summarize the results of our investigation of the mechanism of evolution in the early stages, using the terminology of general cybernetic concepts. Having begun to think in this direction, we shall easily detect one general characteristic of transitions from lower to higher stages: In each stage the biological system has a subsystem which may be called the highest controlling device; this is the subsystem which originated most recently and has the highest level of organization. The transition to the next stage occurs by multiplication of such systems (multiple replication) and integration of them—by joining them into a single whole with the formation (by the trial and error method) of a control system headed by a new subsystem, which now becomes the highest controlling device in the new stage of evolution. We shall call the system made up of control subsystem X and the many homogeneous subsystems A_1, A_2, A_3 . . . controlled by it a *metasystem* in relation to systems A_1, A_2, A_3 . . . . Therefore we shall call the transition from one stage to the next the *metasystem transition* (see figure 3.1). This concept will play a crucial part in our subsequent presentation. The metasystem transition creates a higher

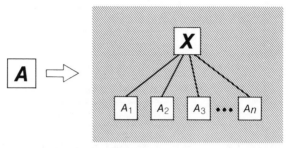

Figure 3.1. The metasystem transition.

level of organization, the *metalevel* in relation to the level of organization of the subsystems being integrated.

From the functional point of view the metasystem transition is the case where the activity α, which is characteristic of the top control system at a lower stage, becomes controlled at the higher stage and there appears a qualitatively new, higher, type of activity β which controls the activity α. Replication and selection bring about the creation of the necessary structures.

The first metasystem transition we discern in the history of animals is the appearance of movement. The integrated subsystems are the parts of the cell that ensure metabolism and reproduction. The position of these parts in space is random and uncontrolled until, at a certain time, there appear organs that connect separate parts of the cell and put them into motion: cell membranes, cilia, flagella. A metasystem transition occurs which may be defined by the formula:

$$\text{control of } position = movement.$$

In this stage movement is uncontrolled and not correlated in any way with the state of the environment. Nature's next task is to control it. To control motion means to make it a definite function of the state of the environment. This leads to irritability. Irritability occurs when—under the influence of external factors—there is a change in the state of some segments of the cell, and when this change spreads to other sectors—specifically those which ensure movement. Thus, the formula for the metasystem transition from the second stage to the third is:

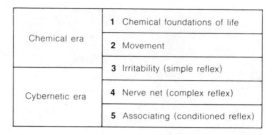

Chemical era	**1** Chemical foundations of life
	2 Movement
Cybernetic era	**3** Irritability (simple reflex)
	4 Nerve net (complex reflex)
	5 Associating (conditioned reflex)

Figure 3.2. Stages in the evolution of life before the era of reason.

control of *movement = irritability*.

The integration of cells with formation of the multicellular organism is also a transition from a system to a metasystem. But this transition concerns the structural aspect exclusively and is not describable in functional terms. From a functional point of view it is ultimately unimportant whether reproduction and integration of a certain *part* of the organism occur or whether organisms are integrated as whole units. This is a technical question, so to speak. Irritability is already manifested in unicellular organisms, but it reveals its capabilities fully after cell integration.

An important characteristic of the metasystem transition must be pointed out here. When the subsystems being integrated are joined into a metasystem, specialization occurs; the subsystems become adapted to a particular activity and lose their capability for other types of activity. Specialization is seen particularly clearly where whole organisms are integrated. Each subsystem being integrated in this case contains a great deal which is "superfluous"—functions necessary for independent life but useless in the community, where other subsystems perform these functions. Thus, specialized muscle and nerve cells appear in the multicellular organism.

In general we must note that the integration of subsystems is by no means the end of their evolutionary development. We must not imagine that systems $A_1, A_2, A_3, \ldots$ are reproduced in large numbers after which the control device X suddenly arises "above them." On the contrary, the rudiments of the control system form when the number of subsystems A_i is still quite small. As we saw above, this is the only way the trial and error method can operate. But after control subsystem X has formed, there is a massive replication of subsystems A_i and during this process both A_i and X are refined. The appearance of the structure for control of subsystems A_i does not conclude rapid growth in the number of subsystems A_i; rather, it precedes and causes this growth because it makes multiplication of A_i useful to the organism. The carrier of a definite level of organization branches out only after the new, higher level begins to form. This characteristic can be called the law of *branching growth of the penultimate level*. In the phenomenological functional description, therefore, the metasystem transition does not appear immediately after the establishment of a new level; it appears somewhat later, after the penultimate level has branched out. The metasystem transition always involves two levels of organization.

Let us continue our survey of the stages of evolution. We shall apply the principle of the metasystem transition to the level of irritability. At this level, stimulation of certain sectors of a unicellular organism or a specialized nerve cell in a multicellular organism occurs directly from the external environment, and this stimulation causes direct (one-to-one) stimulation of muscular activity. What can control of irritability signify? Apparently, creation of a nerve net whose elements, specifically the effectors, are not stimulated by the environment directly but rather through the mediation of a complex control system. This is the stage of evolution we related to the concept of the *complex reflex*. The control of irritability in this stage is seen especially clearly in the fact that where there is a goal, stimulation of the effectors depends not only on the state of the environment but also upon this goal—that is, on the state of certain internal neurons of the net. Thus, the formula for this metasystem transition (from the third stage to the fourth) is:

$$\text{control of } \textit{irritability} = \textit{complex reflex}.$$

What next?

■ CONTROL OF THE REFLEX

NO MATTER how highly refined the nerve net built on the principle of the complex reflex may be, it has one fundamental shortcoming: the invariability of its functioning over time. The animal with such a nervous system cannot extract anything from its experience; its reactions will always be the same and its actions will always be executed according to the same plan. If the animal is to be able to learn, its nervous system must contain some variable components which ensure change in the relations among situations and actions. These components will therefore carry out control of reflexes. It is commonly known that animals have the ability to learn and develop new reflexes. In the terminology introduced by I. P. Pavlov, the inborn reflex included in the nervous system by nature is called an *unconditioned reflex* while a reflex developed under the influence of the environment is called a *conditioned reflex*. When we speak of a complex reflex we have in mind, of course, an unconditioned complex reflex. The presence of components that control complex reflexes manifests itself, in experiments with teaching animals, as the ability to form conditioned reflexes.

We cannot, however, equate the concept of the conditioned reflex with the concept of control of a reflex. The latter concept is broader. After all, our concept of the complex reflex, taken in the context of the description of general principles of the evolution of the nervous system, essentially signifies any fixed connection between the states of classifiers, representation fixers, and effectors. Therefore, control of reflexes must be understood as the creation, growing out of individual experience, of any variable connections among these objects. Such connections are called *associations of representations* or simply *associations*. The term "representation" here is understood in the broad sense as the state of any subsystems of the brain, in particular the classifiers and effectors. We shall call the formation of associations *associating* (this terminology is somewhat awkward, but it is precise). Thus, the fifth stage of evolution is the stage of associations. The formula for the metasystem transition to this stage is:

control of *reflexes* = *associating*.

■ THE REFLEX AS A FUNCTIONAL CONCEPT

THE CONCEPTS of the reflex and the association are *functional,* not *structural* concepts. The connection between stimulus *S* and response *R* in the reflex (see figure 3.3) does not represent the transmission of information from one subsystem to another, it is a transition from one generalized state to another. This distinction is essential to avoid confusing the reflex, as a definite functional diagram which describes behavior, with the embodiment of this diagram, that is, with the cybernetic device that reveals this diagram of behavior.

Confusion can easily arise, because the simplest embodiment of reflex behavior has a structural diagram that coincides externally with the diagram shown in figure 3.3, except that *S* and *R* in it must be understood as physical subsystems that fix the stimulus and response. This coincidence is not entirely accidental. As we have already said in defining the functional diagram, breaking the set of all states of the system down into subsets which are ascribed to vertices of the graph is closely tied to breaking the system down into subsystems. Specifically, each subsystem that can be in two states (yes/no) can be related to the set of all states of the system as a whole for which this system is in a definite state, for example "yes." More simply, when defining the generalized state we consider only the state of the given subsystem, paying no attention to what is happening with the other subsystems. Let us assume that the letters *S* and *R* signify precisely these subsystems, that is to say, subsystem *S* is the discriminator for stimulus (set of situations) *S* and subsystem *R* is the effector that evokes response *R*. Then the statement that "yes" in subsystem *S* is transmitted along a communications channel (arrow) to subsystem *R*, put-

Figure 3.3. Functional diagram of the unconditioned reflex.

ting it also in the "yes" state, coincides with the statement that the generalized state S switches (arrow) to state R. Thus the structural and functional diagrams are very similar. It is true that the structural diagram in no way reflects the fact that "yes" evokes a "yes" not a "no," whereas this is the very essence of the reflex. As we have already said, the reflex is a functional concept.

■ WHY ASSOCIATIONS OF REPRESENTATIONS ARE NEEDED

THESE PRELIMINARY considerations were required in order for us to be able to better grasp the concept of association and the connection between a functional description using associations and a structural description by means of classifiers. Because each classifier can be connected to one or several generalized states, there is a hierarchy of generalized states corresponding to the hierarchy of classifiers. When introducing the concept of the classifier we pointed out that for each state of the classifier (we can now say for each *generalized state* of the system as a whole) there is a corresponding, definite concept at the input of the system—that is, the input situation is affiliated with a definite set. The concepts of the Aristotelian "concept" and the "generalized state" are close to one another; both are sets of states. But the "generalized state" is a more general concept and may take account of the state not just of receptors but also of any other subsystems, in particular classifiers. This is essential to follow the dynamics of the state of the system during the process of information processing.

Let us see how the generalized states of the K level of the hierarchy and the next level, $K + 1$, are interconnected. As we know, the chief task of the classifiers is to store "significant" information and discard "insignificant" information. This means that there is some set of states on level K which in the functional diagram has an arrow going from each of the states to the same state at level $K + 1$. In figure 3.4 below, the representations (generalized states) T_1 and T_2 evoke representation U equally. If T_1 and T_2 always accompany one another this diagram will unquestionably be advantageous to the animal. He does not have to know that T_1 and $T12$ are occurring; it is enough if he knows that U is occurring. In this way superfluous information is discarded and useful information is compressed. The compression of information is possible because T_1 and T_2 are always encountered together. This is a fact which is external to the nervous system and refers only to the stream of situations being fed to it. It testifies to the existence of a definite organization in the stream of situations, which is a consequence of the organized nature of the environment surrounding the animal. The organization of the nervous system and its activity (the system of reflexes) reflect characteristics of the environment. This happens because, by testing different ways to discard information, nature finally finds the variation where the information discarded is indeed superfluous and unnecessary owing to the partially organized nature of the environment.

In the stage of the unconditioned reflex the structure of such connections, as shown in figure 3.4, does not change during the life of the animal and is the same for all animals of the given species. As we have already said, however, such a situation is not satisfactory. The metasystem transition occurs, and the connections between generalized states become controlled. Now if T_1 and T_2 in the individual experience of the animal always (or at least quite often) accompany one another, new connections form in the animal brain which are not determined uniquely by heredity. This is *associating*—the formation of a new association of representations.

It is clear that associations form among representations of the highest level of the hierarchy. Thus, the most general correlations in the environment, those which are the same for all times and all places of habitation, are reflected in the permanent organization of the lower

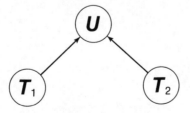

Figure 3.4. Association of representations.

levels of classifiers. The more particular correlations are reflected by variable connections at the highest level.

■ EVOCATION BY COMPLEMENT

THE DIAGRAM shown in figure 3.4 may cause misunderstanding. When speaking of an association of representations we usually mean something like a two-way connection between T_1 and T_2, where T_1 evokes T_2 and T_2 evokes T_1. But in our diagram both representations evoke something different, specifically U, and there are no feedback arrows from U to T_1 and T_2.

In fact, the diagram shown in figure 3.4 more closely corresponds to the concept of the association of representations than a diagram with feedback does. Specifically, it contains an evocation, in a certain sense, of representation T_2 by representation T_1 (and vice versa), but this is evocation by complement. The representation U contains both T_2 and T_1; after all, it was conceived by our nervous system as equivalent to the simultaneous presence of T_1 and T_2. Therefore, when T_1 evokes U in the absence of T_2, then T_2 is contained concealed in U itself. By evoking U we, so to speak, complement T_1 with the nonexistent T_2.

This process of mental complementing is in no way related to the fact that the association is developed by learning. Only the method by which the brain processes information plays a part here. When inborn lower-level mechanisms operate the effect of the complementing shows itself even more clearly; no kind of learning or training will weaken or strengthen it.

Look at figure 3.5. In it you see not just points but also a line,

Figure 3.5. The points make a line.

an arc. In fact there is no line at all. But you mentally supplement (complement) the drawing with points so that a solid line is formed. In terms of figure 3.4, T_1 here is the actually existing points, U is the line, and T_2 is the complementary points. The fact that you discern a nonexistent line testifies to the presence in the brain (or in the retina) of classifiers which create the representation of U.

Why did these classifiers arise? Because the situations arriving at the input of our visual apparatus possess the characteristic of continuity. The illuminations of neighboring receptors of the retina are strongly correlated. The image on the retina is not a mosaic set of points, it is a set of light spots. Therefore, translating the image into the language of spots, the brain (we say "brain" arbitrarily, not going into the question of where the translation is in fact made) rejects useless information and stores useful information. Because "consisting of spots" is a universal characteristic of images on the retina, the language of spots must be located at one of the lowest levels and it must be inborn. The line which we "see" in figure 3.5 is a long, narrow spot.

■ SPOTS AND LINES

NOTICE, we have reduced the concept of the line to the concept of the spot. We had to do this because we were establishing the theoretical basis for the existence of the corresponding classifiers. In reality, it is possible to conclude from the two-dimensional continuity of the image on the retina that the basic concept for the brain should be the concept of the spot, not the line. The line can be included as either an exotically shaped spot or as the boundary between spots. This theoretical consideration is confirmed by numerous observations.

A circle formed by the vertices of angles is clearly seen in figure 3.6a. In figure 3.6b the vertices of the angles are located at exactly the same points, but their sides are directed every which way, some outside and some inside. As a result the circle disappears. It is possible to follow the vertices along, switching attention from one to another, and ascertain that they are set in a circle, but you cannot *see* this as you can in the first drawing, even though the points which

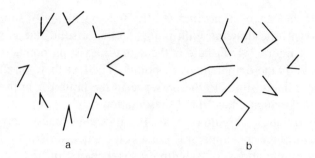

a   b

Figure 3.6. Concealed circle formed by the vertices of angles.

make up the circle are all vertices of angles and all lie on the circumference of the circle. The simplest machine program for recognizing cirles would "see" the circle in figure 3.6b (as well as figure 3.6a). But our eye does not see it. In Figure 3.6a, where all the rays are directed out, however, our eye glosses them into something like a rim and clearly sees the internal circle, a two-dimensional formation, a spot. The circumference, the boundary of this spot, also becomes visible.

There are many visual illusions resulting from the fact that we "see in spots." They offer instructive examples of inborn associations. One of the best ones is shown in figure 3.7. Figure 3.7a is a square, and its diagonals intersect at right angles. Figure 3.7b is constructed of arcs, but its vertices form precisely the same square as in figure 3.7a, and therefore the diagonals also intersect at right angles. This is almost impossible to believe, so great is the illusion that the diagonals of figure 3.7b are approximated to the vertical. This illusion may be explained by the fact that alongside the microcharacteristics of the figure—that is, the details of its shape—we always per-

a   b   c

Figure 3.7. The illusion of the approximation of diagonals.

ceive its macrocharacteristics, its overall appearance. The overall appearance of figure 3.7b is that of a spot which is elongated on the vertical. The degree of elongation may be judged by figure 3.7c. This figure is a rectangle whose area is equal to the area of figures 3.7a and b, while the ratio of its width to its height is equal to the ratio of the average width of figure 3.7b to its average height. The hypothetical classifier which records the overall elongation of the figure will arrive at the same state in contemplating figure 3.7b as in contemplating figure 3.7c. In other words, whether we desire it or not, figure 3.7b is associated in our mind with the rectangle in figure 3.7c. Following the diagonals in figure 3.7b in our mind, we equate them with the diagonals of figure 3.7c, which form acute vertical angles. The classifier that records elongation of the spot is unquestionably a useful thing; it was especially useful for our distant ancestors who did not perceive the world in more subtle concepts. But because we cannot switch it on or off at will, it sometimes does us a disservice, causing visual deception.

■ THE CONDITIONED REFLEX AND LEARNING

BUT LET US RETURN from inborn associations to developed ones, that is, to the actual associating of representations. The very essence of the metasystem transition from the fourth stage of evolution to the fifth lies in the difference between the suffixes of two words from the same root. The *association* is simply one of the aspects of the complex reflex, while *associating* is control of associations: the formation of new associations and disappearance of old ones.

The capability for associating representations appears most fully as the capability for forming (and therefore also recognizing) new concepts. The dog that recognizes its master from a distance may serve as an example.

The Pavlovian conditioned reflex is a more particular manifestation of the capability for associating. The diagram of this reflex is shown in figure 3.8. The unconditioned stimulus S_1 (food) is always accompanied by the conditioned stimulus S_2 (a whistle), and as a result they become associated in one representation U, which, be-

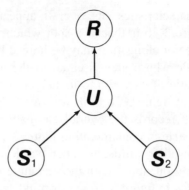

Figure 3.8. Diagram of the conditioned reflex.

cause of the presence of S_1 in it, causes the response R (salivation). Then stimulus S_2 causes U, and therefore R, even where S_1 is not present. The whistle causes salivation.

A question may arise here. The conditioned reflex arises on the basis of the unconditioned reflex whose diagram is $S{\to}R$. At the same time, if the conditioned stimulus is removed in figure 3.8, we shall obtain the diagram $S_1{\to}U{\to}R$. How do we know that step U exists? Is this an arbitrary hypothesis?

In reality the diagram shown in figure 3.8 contains absolutely no hypotheses. We shall emphasize once more that this diagram is functional, not structural. We are making no assumptions about the organization of the nerve net; we are simply describing observed facts, which are these: first, state S_1 leads, through the mediation of some intermediate states, to state R; second, state S_2 in the end also leads to R. Therefore, at some moment these two processes are combined. We designate the state at this moment U and obtain the diagram we are discussing. In this way our diagram, and our approach in general, differ from the Pavlovian diagram of the reflex arc, which is precisely the structural diagram, a physiological model of higher nervous activity.

The process of learning, if it is not reduced to the development of certain conditioned reflexes (that is, touching only the discriminatory hierarchy) also includes the element of acquiring know-how, development of specific skills. The process of learning also fits within the diagram of associating representations in the general meaning we give to this concept. After all, learning involves the development and reinforcement of a detailed plan to achieve a goal, a new plan that did not exist before. The plan may be represented as an organized group of associations. Let us recall the regulation diagram (figure 2.6). With a fixed goal the comparison block must juxtapose a definite action to each situation. The "untaught" comparison block will test all possible actions and stop at those which yield a reduction in the discrepancy between the situation and the goal (the trial and error method). As a result of learning a connection is established between the situation and the appropriate action (which is, after all, a representation also) so that the "taught" comparison block executes the necessary action quickly and without error.

Now for a few words about *instinct* and the relationship between instinctive behavior and behavior developed through learning. Obviously, instinct is something passed on by inheritance—but exactly what? In the book already referred to, Miller, Galantier, and Pribram define instinct as a "hereditary, invariable, involuntary plan." Plans, as we know, are organized on the hierarchical principle. It is theoretically possible to assume the existence of an instinct that applies to all stages of the hierarchy, including both the general strategy and particular tactical procedures all the way to contracting individual muscles. "But if such an instinct does exist," these authors write, "we have never heard of it." The instinct always keeps a definite level in the hierarchy of behavior, permitting the animal to build the missing components at lower levels through learning. A wolf cub which is trying to capture a fleeing animal unquestionably acts under the influence of instinct. But it is one thing to try and another to succeed. "It may be considered," Miller, Galantier, and Pribram write, "that copulation is an instinctive form of behavior in rats. In certain respects this is in fact true. But the crudeness of copulative behavior by a rat which does not have experience in the area of courting shows plainly that some practice in these instinctive responses is essential."

As the organization of an animal becomes more complex and its ability to learn grows in the process of evolution, the instincts "retreat upward,'' becoming increasingly abstract and leaving the animal more and more space for their realization. Thus the behavior of animals becomes increasingly flexible and changes operationally with changes in external conditions. The species' chances for survival grow.

■ MODELING

IN OUR DISCUSSION of associations of representations thus far we have completely ignored their dynamic, temporal aspect; we have considered the representations being connected as static and without any coordinate in time. But the idea of time can enter actively into our representations. We can picture figures that are moving and changing at a certain speed and we can continue the observed process mentally. A wheel rolls down the road. We close our eyes for a second or two and picture the movement of the wheel. Upon opening our eyes we see it in exactly the place where we expected it. This is, of course, the result of an association of representations, but this means an association, or more correctly representations, which are organically bound up with the passage of time. The wheel's position x at moment t is associated with the position x_1 at moment $t + \Delta t$, with position x_2 at moment $t + 2\Delta t$, and so on. Each of these representations includes a representation of the time to which it refers. We do not know the mechanism by which this inclusion is made and, in conformity with our approach, we shall not construct any hypotheses regarding this. We shall simply note that there is nothing particularly surprising in this. It is commonly known that an organism has its own time sensor, the ''internal clock.''

The association of representations that have a time coordinate enables us to foresee future situations in our imagination. We have just established the existence of such representations relying on internal, subjective experience. But the fact that animals also reveal the capability for foresight (look at the way a dog catches a piece of sugar) leads us to conclude that animal representations may also have a time coordinate.

Speaking in the language of cybernetics, the interconnection of representations which have a time coordinate and the resultant capability to foresee the future is simply *modeling,* constructing a *model* of the environment.

Let us give the general concept of the model. We shall consider two systems α and β. Let us assume that to each state A_i of system α we can somehow juxtapose one definite state B_i of system β. The inverse correspondence does not have to be unique (single-valued); that is, many states of α may correspond to one state of β. Because, according to our definition, the *generalized state* is a set of states, this proposition may be described as a one-to-one correspondence of the *states* of system β to the *generalized states* of system α. This is necessary but not sufficient to consider system β a model of system α. Additionally there must be a transformation $T(t)$ of system β which depends on time t and *models* the natural passage of time in system α. This means the following: Suppose that system α is initially in generalized state A_1 which corresponds to state B_1 of system β. Suppose that after the passage of time t the state of system α becomes A_2. Then the conversion $T(t)$ should switch system β to state B_2, which corresponds to generalized state A_2. If this condition is met we call system β a model of system α.

The conversion $T(t)$ may involve nothing more, specifically, than permitting system β by itself to change its state with time. Such models are called *real time* models.

The besiegers dug an opening under the fortress wall and placed several barrels of powder in it. Next to them a candle was burning and from the base of the candle a trail of powder ran to the barrels. When

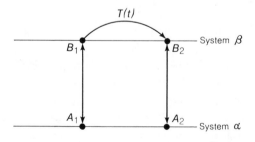

Figure 3.9. Diagram of modeling.

the candle burned down the explosion would take place. An identical candle lighted at the same time was burning on a table in the tent of the leader of the besieging forces. This candle was his model of the first candle. Knowing how much time remained until the explosion he gave his last orders. . . . Wild faces leaned over the table, hairy hands clutched their weapons. The candle burned down and a fearsome explosion shook the air. The model had not let them down.

The image on a television screen when a soccer game is being broadcast may also be formally considered a model of the soccer field and stands. All conditions are in fact observed. But one senses a great difference between the case of the two candles and the case of the soccer broadcast—a difference in the information links between systems α and β. Any image β of object α is a model of it in the broad sense; but there is a continuous flow of information from α to β and it is only thanks to this flow that the correspondence between states α and β is kept. With information access to β, we in fact have access to α. System β operates as simply a phase in the transmission of information from α.

The situation is quite different in the case of the two candles. Candle β burns at the same speed as candle α, but independently of it. The leader of the besieging forces does not have access to candle α and cannot receive any real information regarding its state. By modeling he compensates for this lack and obtains equivalent information. System β here plays a fundamentally different and more significant role. A spatial barrier is overcome, so to speak, by this means and this is done without establishing any new information channels.

Even more important is the case where the model helps overcome a barrier of time rather than space. One cannot, alas, lay an information channel to the future. But a model permits us to operate as if there were such a channel. All that is required is that execution of the conversion $T(t)$ on the model take less time than time t itself. Many other examples could be given of the use of such models in modern life, but that is hardly necessary. Let us return once again to associations of representations.

We have seen that associations of static representations reflect the existence of spatial correlations, interrelationships in the environment. In exactly the same way associations of dynamic representations (models created by the brain) reflect dynamic temporal correlations that characterize the environment. Situation x after time t evokes (or may evoke) situation y—that is the general formula for such correlations, and to the brain these correlations are imprinted in the form of the corresponding associations.

■ COGNITION OF THE WORLD

WHAT IS *knowledge?* From a cybernetic point of view, how can we describe the situation where a person or animal knows something or other? Suppose we know there are two people in an adjacent room. Since they really are there, if we go into the room we shall see two people there. Because we do know this, we can, without actually entering the room, imagine that we are opening the door and entering it; we shall picture the two people who are in the room. In our brains, therefore, an association of representations takes place which enables us to foresee the results of certain actions; that is, there is a certain *model* of reality. For the same reason, when we see a rolling wheel, we know where it will be a second later, and for the same reason, when a stick is shaken at a dog the animal knows that a blow will follow, and so on.

Knowledge is the presence in the brain of a certain model of reality. An increase in knowledge—the emergence of new models of reality in the brain—is the process of *cognition*. Learning about the world is not a human privilege, but one characteristic of all higher animals. The fifth stage of evolution may be called the stage of individual cognition of the world.

CHAPTER FOUR
The Human Being

■ **CONTROL OF ASSOCIATING**

WE HAVE COME to the most exciting moment in the history of life on earth, the appearance of the thinking being, the human being. The logic of our narrative compels us to link the appearance of thought with the next metasystem transition. We still know so little about the process of thinking and the structure of the thinking brain that any theory claiming to explain this phenomenon as a whole is hypothetical. Thus, our conception of thinking must also be treated as a hypothesis. However, this conception indicates the place of thinking in the series of natural phenomena and, as we shall see, puts a vast multitude of facts together in a system. The complete absence of particular, arbitrary assumptions, which ordinarily must be made when a theory includes a structural description of a little-studied object, is another positive feature. The core of our conception is not some hypothesis regarding the concrete structure and working mechanism of the brain, but rather a selection of those functional concepts through which a consistent and sufficiently convincing explanation of the facts we know about thinking becomes possible.

Thus, we assert that the appearance of thinking beings, which marks the beginning of a new stage—perhaps a new era—in evolution (the era of reason) is nothing short of the next metasystem transition, which occurs according to the formula

control of *associating* = *thinking*.

To prove this assertion we shall analyze the consequences that follow from control of associating and equate them with the forms of behavior we observe in thinking beings.

First of all, what is control of associating? Representations X and Y are associated in an animal only when they appear together in its experience. If they do not appear together (as a rule, on many occasions), the association will not arise. The animal is not free to control its associations; it has only those which the environment imposes on it. To control associating a mechanism must be present in the brain which makes it possible to associate any two or several representations that have no tendency at all to be encountered together in experience—in other words, an arbitrary association not imposed by the environment.

This action would appear to be completely meaningless. An elder tree in the garden and an uncle in Kiev—why connect these two totally unrelated facts? Nonetheless, arbitrary associating has profound meaning. It really would be meaningless if brain activity amounted to nothing more than passively receiving impressions, sorting them, grouping them, and so on. But the brain also has another function—its basic one: to control the organism, carrying out active behavior which changes the environment and creates new experience. You can bet that the alarm clock and the holder for the teapot are in no way associated in your consciousness. Nor in the consciousness of your three-year-old son. However, this is only for a certain time. One fine day, for some reason an association between these two objects occurs in the head of the young citizen and he is overcome by an insurmountable desire to rap the alarm clock with the holder. As a result, the objects enter a state of real, physical interaction.

In the metasystem transition, some thing that was once fixed and uniquely determined by external conditions becomes variable and subject to the action of the trial and error method. Control of associating, like every metasystem transition, is a revolutionary step of the highest order directed against slavish obedience by the organism to environmental dictatorship. As is always true in the trial and error method, only a small proportion of the arbitrary associations prove useful and are reinforced, but these are associations which could not have arisen directly under the influence of the environment. And they are what permits a reasoning being those forms of behavior which are inaccessible to the animal that was frozen in the preceding stage.

■ PLAY

HIGHER ANIMALS reveal one interesting form of behavior, play, which relates them to human beings and is a kind of herald of the era of reason. We are not referring to behavior related to mating (which is also called play sometimes), but rather to "pure" and, by appearance, completely purposeless play—play for pleasure. That is how a cat plays with a piece of paper, and how the young (and the adults) of all mammals play with one another.

But what is play? How does this phenomenon arise in the animal world? Play is usually explained as a result of the need to exercise the muscles and nervous system, and it certainly is useful for this purpose. But it is not enough to point out the usefulness of a form of behavior; we still must explain how it becomes possible. When a kitten plays with a piece of paper tied to a string it behaves as if it thinks the paper is prey. But we would underestimate the mental capabilities of the kitten if we supposed that it was actually deceived—it is not. It has caught this paper many times, bitten it, and smelled its offensive, inedible odor. The kitten's representation of the paper is not included in the concept of "prey." However, this representation partially activates the very same plan of action normally activated by the concept of "prey." Similarly, a wolf frolicking with another wolf does not take its playmate for an enemy, but up to a certain point it behaves exactly as if it did. This is the very essence of play. It can be understood as the arbitrary establishment of an association between two objects such as the paper and the prey or the fellow wolf and the enemy. As a result there arises a new representation which, strictly speaking, has no equivalent in reality. We call it a "fantasy," the result of "imagination." Thus the paper plainly is not prey, but at the same time seems to be prey; thus fellow wolf is simultaneously friend and enemy. The synthetic representation generates a synthetic plan of action—a play plan. The wolf is completely serious and tries as hard as it can to overtake and catch its friend, but when it bites it is no longer serious.

Indeed, play exercises the muscles and develops skills which are useful during serious activities, but this impresses one more as a useful side effect than as a special strategic goal for whose sake play forms of behavior are developed. Everyone knows how children love to play, but what is attractive in their play is not so much the pleasure gained from physical exercise or showing one's agility as it is the game as such. When boys play soldiers and girls play dolls they are not exercising anything except their imagination—that is, their ability to make arbitrary associations. It is these arbitrary associations which give them pleasure. Children's play is a phase of development through which every person must inevitably pass in order to become a person. In his remarkable book *From Two to Five,* K. Chukovsky devotes many pages to developing the idea of the absolute necessity for elements of play and fantasy in a child's upbringing. Children cannot get along without these things. They give themselves up to play completely, feeling it to be something needed, important, and serious.

K. Chukovsky writes:

I knew a little boy who was pretending to be a chimney sweep and shouted, "Don't touch me, mama, you'll get dirty! . . ."

Another little boy who had been pretending he was a meatball for quite some time and was conscientiously sizzling in the frying pan pushed his mother away in irritation when she ran up and began kissing him. "How dare you kiss me when I'm cooked!" he shouted.

No sooner had my three-year old daughter Mura spread out her books on the floor in play than the books became a river where she caught fish and washed clothing. Accidentally she stepped on one of the books and exclaimed "Oh, I got my foot wet!" It was so natural that for a minute I believed that the books were water and almost ran over to her with a towel.

In all of these games the children are both the authors and the performers of fantasies which they embody in play-acting. And the desire to believe in their make-believe is so strong that any attempt to bring them back to reality evokes heated protest.

The apparatus for controlling associations first announces its existence through the need for play. And because it exists it must work; it needs something to do. This is just as natural as the lungs needing air and the stomach food.

■ MAKING TOOLS

BUT LET US leave play behind and pass on to the serious acts of serious adult people. When speaking of the origins of human beings the use and manufacture of tools are pointed out as the first difference between humans and animals. The decisive factor here is, of course, making the tools. Animals can also use objects as tools. The woodpecker finch of the Galapagos Islands uses a spine of cactus or small chip to pick worms out of the bark of a tree. No one who has seen pictures of the skillful way the finch manages the spike held in its beak can fail to agree that this is a clear and very artful use of a tool. The California sea otter lies on its back on the surface of the water, places a flat rock on its chest, and breaks mussels open with it. Monkeys sometimes use sticks and stones. These are very meager examples, but they show that in principle there is nothing impossible in animals using tools. In fact, why can't a plan of action passed on by heredity and reinforced by learning include the selection and use of certain types of objects? Concepts such as "long and sharp" or "round and heavy" are fully accessible to animals. It is obvious that examples such as those given above are rare, because the tools that can be received from nature without special manufacture are very imperfect and in the process of evolution animals have greater success using and refining their natural organs: beaks, claws, and teeth. If the use of tools is to become the rule, not the exception, it is necessary to be able to make them or at least to be able to find suitable objects especially for the particular case.

Suppose that you have to drive a nail but do not have a hammer handy. You look around, seeking a suitable object and spot a bronze bust of Napoleon on the table. You have never before had to drive nails with Napoleons. We can even assume that you have never before driven nails with anything but a real hammer. This will not prevent you from taking the bust and driving the nail. You did not have the association "nail = bust," you created it. In your imagination you compared the nail and the bust of Napoleon, pictured how you could drive the nail with him, and then did so.

■ IMAGINATION, PLANNING, OVERCOMING INSTINCT

IF IN THE BRAIN of the animal there is an association between object X, a tool, and object Y, the object of the action (and, of course, if it is physically possible to execute the action), the animal will be capable of using the tool. But if there is no such association, the animal does not "guess" that it should do this. A dog may be trained to drag bench X in its teeth to fence Y, climb up on the bench, and jump from it over the fence, but if it is not taught this it will not figure it out with its own mind. The dog knows very well that the bench can be moved from place to place. It also knows what opportunities open up when the bench is next to the fence; if you put the bench there it will immediately jump up on it and leap over the fence (assuming that there is some need for it to do so). This means that it is able to foresee the result of the combination of X and Y; it has the corresponding model in its brain. But this model is just dead weight, because the dog cannot picture to itself the combination XY as a goal to strive toward; it does not have the *imagination* for this. It is not enough to know what *will* be, one must also imagine what *can* be. The bare formula which equates thinking with control of associating may be translated into less precise but more figurative language by stating that the human being differs from the animal by the possession of imagination.

Let us construct a very simple model of the working of imagination, using A to designate the situation occuring at a given moment and Z for the situation to be achieved. We shall consider that for the given situation only some other situations are immediately achievable. This will be written by the formula:

$$A \rightarrow (B, C, H, Z)$$

where situations immediately achievable from A are shown in parentheses.

Let us assume that a certain animal (or person) knows which situations are achievable from which others—that is, in its brain there is a series of associations that can be represented in formulas resembling the one above. We shall also consider that for each transition from the given situation to another, directly achievable, one the action which executes it is known. We shall not introduce designations for this, however, so as not to clutter up the formula.

If the brain does possess the exact association shown above, and therefore state Z is achievable from A, the animal will immediately execute the necessary action. Now let us suppose that the brain contains the following group of associations:

$$A \rightarrow (B, C, D)$$
$$B \rightarrow (E, F)$$
$$D \rightarrow (G, H, I, J)$$
$$H \rightarrow (B, C)$$
$$I \rightarrow (B, C, Z)$$

In this table there is no action which would switch A to Z, and therefore the animal given this problem will not be able to solve it. It will either do nothing or flounder in confusion, executing all the actions in the table without any order. But the human being will imagine that he has performed action A in order to understand what situations will become accessible to him in this case. In other words, he will create new associations, which can be written as follows:

$$A \rightarrow B \rightarrow E$$
$$A \rightarrow B \rightarrow F$$

It is true that in the given case these associations will prove useless, but continuing with such attempts the human being will finally find the solution:

$$A \rightarrow D \rightarrow I \rightarrow Z.$$

It is also possible, of course, to approach the problem from goal Z. The main thing is that the table of associations itself does not remain unchanged; it becomes an object of work according to the trial and error method and new lines are added to it. Further, these lines do not appear through the influence of the environment (which determines only the initial list of associations), they result from the functioning of a special mechanism which follows its own rules and laws.

The higher animals also have the rudiments of imagination, which manifest themselves, as has already been noted, particularly in games. Elements of imagination can be clearly discerned in the behavior of the anthropoid apes, which show a resourcefulness dogs and other animals cannot attain. There have been experiments in which an ape has used a support (a cube) to reach a suspended lure, and has even placed one cube on top of another if necessary. With a stick, the ape can push a lure out of a segment of pipe. It can find an appropriate stick, and even split it in half if it is too thick and does not go into the pipe. This can be considered the beginning of toolmaking.

All the same, the boundary is not between the dog and the ape, but between the ape and the human being. At some moment our ancestors' ability to control associating crossed a threshold, beyond which it became an important factor for survival. Then this capability was refined during the course of evolution. The metasystem transition was completed; the human being had become distinct from the world of animals.

Many factors played parts in the process of humanization, above all the organization of the limbs of the man-ape. No matter what wise instructions the brain might give, they would come to naught if it were physically impossible to execute them. On the other hand, the existence of organs capable of executing subtle actions does not by itself give rise to thinking: Insects are physically capable of very complex operations; the limbs of the dinosaurs could, in principle, have served as the starting point for the development of arms; the tentacles of the octopus are more perfect in design than our arms. Unquestionably the leading role is played by the brain. At the same time, the arms of the man-ape and the possibility of having them free when walking fostered a situation where the brain's capability for control of associations became (through the mediation of using and making tools) a factor of decisive importance for survival. Other factors, such as a sharp change in natural conditions, could operate in this same direction. It may be that some other circumstances also played a part.

Clarifying the concrete conditions of the origin of the human being and the role played in this process by various circumstances is a complex and interesting problem on which many scientists are working, but it is not the subject of this book. For us it is enough to know that the combination of conditions necessary for the metasystem transition did come about.

Because the goals which are the most important elements in a plan are representations, the ability to associate representations arbitrarily means the ability to make plans arbitrarily. The human being can decide as follows: first I will do A, then B, then C, and so forth. The corresponding chain of associations arises. The human being can decide that it is absolutely necessary to do X. The association "$X-$necessary" arises. New, concrete plans also occur to the animal constantly, but the mechanism of their occurrence is different. They are always part of a more general (standing higher in the hierarchy) plan and, in the end, a part of instinct. The goals the animal sets are always directed to executing an instinctive plan of actions. The instinct is the supreme judge of animal behavior—its absolute and immutable law. The human being also inherits certain instincts, but thanks to the ability to control associations he can get around them and create plans not governed by instinct and even hostile to it. Unlike the animal, the human being sets his own goals. Where these goals and plans are taken from and what purpose they serve is another matter. We shall take this up when we discuss the human being as a social being. For now, the only thing to keep in mind is that the human brain is organized in a way that makes it possible to go beyond the framework of instinctive behavior.

■ THE INTERNAL TEACHER

THE HUMAN BEING does not by any means perform each operation "through personal imagination"—that is, as if discovering it for the first time. On the contrary, a person (at any rate an adult) does most operations without using imagination; they are routine and customary, and they are regulated by already established associations. The mechanism of such operations does not differ from what we observed in animals, and we call the method by which the necessary associations are developed learning, just as with animals. But the mechanism of learning in humans and animals differs radically.

In the animal, new associations are in a certain sense imposed from outside. For an association to form it must have motivational grounds, be related to a negative or positive emotion. Reinforcement is essential. In other words, teaching takes place only by the "carrot and stick" [in Russian, literally, "knout and cake"]. When a person learns, he himself is taking steps toward learning; but this is not because he knows that "learning is useful." The baby does not know this, but it learns most easily and actively. In the baby, associations "simply form" without any reinforcement. This is the functioning of the mechanism for control of associating, which requires nourishment. If he does not have it the person becomes bored, a negative emotion. There is no need for the teacher to force anything on the child, or upon people in general; the teacher's job is simply to provide nourishment for the imagination. Upon receiving this nourishment a person feels satisfaction. Thus, he himself is always learning inside. This is an active, creative process. Thanks to the metasystem transition, the human being acquired his own internal teacher who is constantly teaching him, driving him with the internal stick and luring him with the internal carrot.

The "internal teacher" is not a fanatic; he takes a realistic approach to his pupil's capabilities. Representations which coincide or are close in time by no means always form stable associations. If they did, it would indicate the existence of absolute memory—that is, total recall. We do not know why we do not have this capability; it may be supposed that the brain's information capacity is simply inadequate. But the existence of people whose capabilities for memorization are substantially greater than average appears to contradict this hypothesis and leads us to conclude that the lack of such capability is more likely the result of some details involving the organization of control of associating. In any case, because there is no absolute memory there must be a criterion for selecting associations. One of the human criteria is the same as found among animals: emotional strain. We memorize things involving emotions first of all. But the human being also has another criterion (which is, by the way, evidence of the existence of control of associating): we can decide to memorize something and as a result in fact do memorize it. Finally, the third and most remarkable criterion is that of novelty. We know that people memorize things new to them and let old things go by ("in one ear and out the other"). But what is the difference between "new" and "old?" After all, strictly speaking no impression is ever repeated. In this sense every impression is a new one. But when we hear talk on a hackneyed subject or see hackneyed situations on a movie screen we start to yawn and wave our hand in annoyance: "This is old!" When the stream of impressions fits into already existing models, our "internal teacher" sees no need to change the model and the impressions slip by without any consequences. This is the case when we know ahead of time what is coming. But when we do not know what is coming (or even more so when it contradicts the model) then new associations appear and the model becomes more complicated. The relationship to the model already existing in the brain is the criterion for the novelty of an impression.

As we begin to talk about memory and other aspects of the human psyche, we touch on many unresolved problems. Fortunately, a systematic presentation of human psychology, particularly in its "cyberneticized" variation, is not part of our task. We shall be content with a quick survey of the psychological characteristics that distinguish human beings from animals in order to be sure that they are the natural results of the metasystem transition—the appearance of an apparatus for controlling associating.

We have seen that the control of associating leads to a qualitative difference between the human and the animal in susceptibility to learning. We shall also note in passing that the enormous quantitative difference that exists between these levels for humans and animals, and that is expressed simply in the quantity of information memorized in the process of learning, is also a direct consequence of the metasystem transition. It follows from the aforementioned law of branching of the penultimate level. In this case the penultimate level consists of the physical devices for the formation of associations. Multiplication of these devices means enlarging memory. We shall

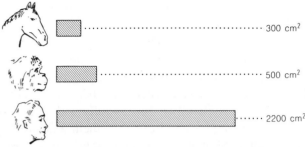

Figure 4.1. Area of the cerebral cortex in the horse, orangutan, and human being.

deviate here from our principle of not considering structural models of the brain to point out the branching of the human cerebral cortex, which according to general (and well-founded) opinion is the storage place for associations.

■ THE FUNNY AND THE BEAUTIFUL

ALL THE SAME, qualitative differences are more interesting. We have already established that the existence of a special apparatus for the control of associating makes learning an active process involving positive and negative emotions for the human being. These are truly human emotions which are inaccessible to beings that do not possess this apparatus. The goal of associating is the construction of a model (or models) of the environment, and we may therefore conclude that a new emotion will be positive if it establishes an association which improves the brain model of the world. This emotion can be called the *pleasure of novelty*, using the term ''novelty'' in the sense we gave it above. The corresponding negative emotion is called *boredom*. We have already enumerated the criteria for establishing and reinforcing associations and separated the criterion of novelty from the criterion of the existence of emotional reinforcement. We had in mind ordinary emotions common to humans and animals. When we raise the pleasure of novelty to the rank of an emotion we can declare the third criterion to be a particular case of the first. Then we can say that involuntary associating always involves emotional reinforcement, but compared to the animal the human being possesses a fundamentally new class of emotions.

Yes, that is right, a class. The "pleasure of novelty" is a very general term which covers a whole class of emotions. We can immediately point out two plainly different representatives of this class: the sense of the funny and the sense of the beautiful. Hardly anyone today would try to maintain that he has fully and finally understood the nature of these emotions and can give them a detailed cybernetic interpretation. Unquestionably, however, they are inseparable from cognition of the world, from the creation of new models.

What makes us laugh? A disruption of the "normal" course of events which is completely unexpected but at the same time natural, and in hindsight entirely understandable: an unexpected association, meaningless at first glance but reflecting some deep-seated relationships among things. All this, of course, creates a new model of the world and gives pleasure proportional to its novelty. When it is no longer new it is no longer funny. When someone tries to make us laugh using a very familiar model we call it "flat" humor. But this is a very relative concept. Everyone is familiar with the situation where a joke is told and one listener bursts out laughing while the other smiles sourly. The difference between the two listeners is obviously the absence or presence of the corresponding model. Another situation very important for clarifying the nature of humor occurs when one person laughs and another glances around uncomprehending. "He didn't get it," they say in such cases. The joke was too subtle for this person; it relied on associations he did not have. What is funny is always on the borderline between the commonplace the unintelligible. Every person has his own borderline and the line shifts in the process of individual development. Nothing shows the level of a person's sophistication so clearly as his understanding of what is funny.

There are more individual differences among people in their sense of the beautiful—a sense more subtle and mysterious than the sense of humor. But here too we find the same dynamism related to the novelty of the impression. Frequent repetition of a pleasing piece of music creates indifference to it, and finally revulsion. A sharp sensation of the beautiful is short in duration; it includes the element of revelation, enchanting surprise. It can also be described as the sudden discernment of some deep order, correspondence, or meaning. If we attempt to interpret this phenomenon cybernetically, we may assume that the sense of the beautiful evokes impressions which give nourishment to the most complex and subtle models, which employ classifiers on the highest level. These classifiers must, of course, compress information to the maximum degree and recognize extremely complex concepts. That is what discernment of a deep internal order in apparent disorder is.

All models are hierarchical. The more complex is built on the simpler, and the higher rests on the lower. A person may be insufficiently developed in esthetic terms and not see beauty in a place where others do see it. To an untrained listener a masterpiece of symphonic music will seem to be a meaningless cacophony. On the other hand, a banal melody or a primitive geometric ornament will not elicit a sensation of the beautiful in us; in this case the order is too obvious. When we say in ''us,'' we are speaking of modern, civilized people. It is possible that a Neanderthal would be shaken to the depth of his soul upon seeing a series of precisely drawn concentric circles. The beautiful too is always found on the borderline between the commonplace and the unintelligible. Shifting this line, which we can define as esthetic education, is cognition of the world and the consequent construction of new models in the brain.

We are taking the sense of the beautiful in its pure form. In reality it is ordinarily bound up with other human feelings, often forming inseparable groups and therefore influencing many spheres and aspects of societal life. The value of esthetic experiences, which may be called its applied value, has long been widely recognized. The situation with pure esthetics is worse. Now and again through the course of human history there have been calls to put an end to pure esthetics once and for all, as something not simply useless but even directly harmful. (The harm has been understood in different ways. Some have proclaimed beauty to be sinful while others have argued that it distracts from the class struggle.) On the other hand, there have been attempts by the vulgar materialist school to explain and "justify" the beautiful by reducing it to the useful in the most ordinary, everyday sense of the word. This is like someone praising a transistor radio because it can be used to drive nails and crack nuts. This attitude arises from a failure to understand that pure esthetic education trains the brain to perform its highest and most subtle functions. The brain is unitary. The models created in the process of esthetic education unquestionably influence a person's perception of the world and his creative activity. Exactly how this happens is unknown. Esthetic education is more precious because we know of no substitute for it.

■ LANGUAGE

UNTIL NOW we have considered the human being as an individual only and have been interested in the capabilities of the human brain. With this approach it is not at all obvious that the appearance of the human being on earth was such a major revolution in the history of life. The frog was more intelligent than the jellyfish. The dog was more intelligent than the frog. The ape was more intelligent than the dog. Now there appeared a being which was more intelligent than the ape. Well, so what?

The revolution was created by the appearance of human society which possessed a definite culture, above all language. The key aspect here is language. *Language* in general is understood to be a certain way of correlating objects R_i, which are considered to be some kind of primary reality, to objects L_i, which are called the *names* of objects R_i and are viewed as something secondary, especially created to be correlated to objects R_i. In relation to the name L_i object R_i is called its *meaning*. The aggregate of all objects L_i is frequently also called a language (in a more expanded form it would be better to call it the material fixer or carrier of the language). The set of objects R_i can be much broader and more varied than the set of language objects L_i. This is the case, for example, with natural languages such as Russian, English, and others. It is clear that an enormous amount of information is lost when word descriptions are substituted for the perception of real objects and situations. In those cases where the information levels of objects R_i and L_i are on the same order of magnitude, the cybernetic term *code* is often used in place of the word "language." The transition from R to L is called *coding* and the opposite transition from L to R is *decoding*. Thus, when a message is transmitted in 'Morse' Code by radio, the initial text—a set of letters—is coded in a set of dots and dashes. In this code (language), information travels through the air and is received at an assigned point, where decoding from the language of dots and dashes to the language of letters takes place. In this case the process of coding and decoding does not cause information loss.

Because there are no more convenient and generally accepted terms than coding and decoding for the transition from the meaning to the name we shall use these terms in the most general sense, disregarding the ratio of information levels (and calling language simply that, and not "code").

Objects R_i and L_i may be arbitrary in nature; they do not have to be physical objects but, speaking generally, may be phenomena such as sound oscillations. Let us note that "phenomenon" is the most general term we can use to designate a part of physical reality which is limited in space and time; "physical object," by contrast, is a less clear-cut concept which refers to phenomena of a special type that reveal a certain stability: they have a surface across which the exchange of matter does not take place. This concept is not clear-cut because there are no absolutely impenetrable boundaries in reality and so-called "physical objects" are continuously changing. This is a relative concept which only reflects a low rate of change.

Elementary language is also found among animals, especially among those living in communities, which therefore must somehow coordinate their actions and "clarify relationships." We call this language elementary only in comparison with human language; by itself animal language is not at all simple and evidently well satisfies the needs of members of the community for the exchange of information. The danger signal, the call for help, the intention to initiate mating relationships and the acceptance or rejection of this intention, the order to obey, and the order for everyone to head home—these and other components are found in the languages of most birds and mammals.

They are expressed by gestures and sounds. When bees return to the hive from a honey-gathering expedition they show the other bees where they have been by performing certain unique movements which resemble a dance.

■ CREATION OF LANGUAGE

HUMAN LANGUAGE differs radically from animal language. As was the case with the use of tools, the animal language is something given at the start—an element of instinctive behavior. If language does change it is only along with changes in behavior accompanying the general evolution of the species. For the human being, language is something incomparably more mobile and variable than behavior. The human being himself creates language; he has the capability (and even the need) to assign names, something no animal can do. Giving names to phenomena (specifically, to physical objects) is perhaps the simplest and most graphic manifestation of the control of associating. There is nothing in common between the word "lion" and a real lion, but nonetheless the association between the word and its meaning is established. It is true that many onomatopoetic words appeared in the dawn of human culture. There is an abundance of such words in the languages of primitive cultures. The same thing is even more true of gestures, which have obviously always been imitative at base. But this does not change the nature of the association between the name and the meaning as the result of deliberate associating. Let us suppose that in some primitive language the lion is called "rrrrr." The association between "rrrrr" and the lion does not arise because this sound can be confused with the lion's roar (it would be quite a hunter who was capable of making such a mistake), but because in searching for a *name* for the lion the human being sorts through the animal's characteristics in his imagination and selects one of them which permits at least an approximate reproduction. The creator of a name perceives it subjectively as something close to the meaning— something like it or, to be more precise, likened to it. This is because the objective resemblance between the name and the meaning cannot be large; it is almost nil and serves only as an umbilical cord, which withers away soon after the name is born. The association between the name and the meaning does not arise at all in the way that the association between types of dishes and salivation arose in Pavlov's experiments with dogs. The latter was a conditioned reflex, but the former is creation of language. The occasion that brought about the choice of the name is forgotten and the name itself is transformed, but the relationship between the name and its meaning does not suffer from this.

■ LANGUAGE AS A MEANS OF MODELING

LANGUAGE ARISES as a means of relationship, of communication among members of a primitive community. But once it has arisen it immediately becomes a source of other, completely new possibilities which are not in principle related to relationships among people. What these possibilities are we shall demonstrate with the example of the language of numbers.

Let us imagine a young man from the primitive Nyam-Nyam tribe. We shall call him Uu. Now let us see how he performs the duties of scout.

Uu is lying behind a thick old oak tree and keeping constant watch on the entrance to a cave on the opposite bank of the river. At sunrise a group of men from the hostile Mayn-Mayn tribe approach. They are obviously planning something bad, probably setting up an ambush in the cave. They scurry back and forth, now going in the cave and now coming out, first disappearing in the forest and then returning to the cave. Each time an enemy enters the cave Uu bends over one finger, and when an enemy comes out of the cave he unbends one finger. When the enemy goes away Uu will know if they have left an ambush party and, if they have, how many people are in it. Uu will run to his own tribe and tell them with his fingers how many enemy men remain in the cave.

Our hero is able to know how many men are in the cave at any moment because he has used his fingers to construct a *model* of that part of the external world which interests him. And what interests him is the cave and the enemy in it. In his model one bent finger corresponds to each enemy in the cave. A bent finger is the *name* of the enemy in the cave; an enemy in the cave is the *meaning* of a bent finger. The operations performed on the names, bending and unbending the fingers, correspond to the entrances and exits of enemies from the cave. This is a language. It can be called a finger language if we are looking at the physical material from which the model is constructed or a number language if we are looking at the method of correlating names with meanings. And this language is used not so much for information transmission as for constructing a model which is needed precisely as a model—as a means of foreseeing events, a means of finding out indirectly that which cannot be found out directly. If his native Nyam-Nyam tribe is far away and Uu does not intend to tell anyone how many enemy men are in the cave, in order to plan his own course of action he still has reason to count them. The communicative use of language (a means of communication among people) is supplemented by the noncommunicative use of language (a means of constructing models of reality).

We come now to the crux of the matter. The modeling function of language is that final element which we lacked for assessing the appearance of the human being on earth as the boundary between two ages, as an event of cosmic importance. When an astronomer determines the position of the planets in the sky, makes certain calculations, and as a result predicts where the planets will be after a certain interval of time, he is essentially doing the same thing that Uu did when he bent and unbent his fingers as he watched the entrance to the cave. Art, philosophy, and science—all these are simply the creation of linguistic *models of reality*. The remainder of this book will be devoted to an analysis of this process, its laws and results. But first we shall take a general look at its place in the evolution of the universe.

■ SELF-KNOWLEDGE

THE ANIMAL has no concept of itself; it does not need this concept to process information received from outside. The animal brain can be compared to a mirror that reflects the surrounding reality but is not itself reflected in anything. In the most primitive human society each person is given a name. In this way, a person, represented in the form of sentences containing the person's name, becomes an object for his or her own attention and study. Language is a kind of second mirror in which the entire world, including each individual, is reflected and in which each individual can see (more correctly, cannot help but see!) his or her own self. Thus the concept of "I" arises. If the stage of cognition may be called the concluding stage of the cybernetic period, the era of reason is the era of self-knowledge. The system of two mirrors, the brain and language, creates the possibility of a vast multitude of mutual reflections without going outside the space between the mirrors. This gives rise to the unsolved riddles of self-knowledge, above all the riddle of death.

■ A CONTINUATION OF THE BRAIN

LET US SUPPOSE that three enemies enter the cave and two come out. In this case even without the use of fingers the primitive man will know that one enemy has remained in the cave. A model he has in his brain is operating here. But what if 25 go in and 13 or 14 come out? In this case the human brain will be impotent; it does not contain the necessary model, the necessary concepts. We instantaneously and without error distinguish sets of one, two, three, and four objects and can imagine them clearly. These concepts are given to us from nature and are recognized by the nerve net of the brain, just as the concepts of spot, line, contiguity, and the like are. It is not so easy with concepts expressed by the numbers between five and eight; here a great deal depends on individual characteristics and training. As for the concepts of "nine," "ten," and so on, with very rare exceptions they are all merged into the single concept of "many." And then the human being creates a language whose material carrier (for example the fingers) serves as a fixer of new concepts, performing the functions of those classifiers for which no room was found in the brain. If there are not enough fingers then pebbles, little sticks, and chips will come into play . . . and in the more developed languages, numbers and sets of numbers. The language used is not important; the ability to encode is. The process of counting serves for the recognition of new concepts, performing the functions of the nerve net which is put in a stimulated state by some particular classifier. As a result of counting, the object R (which for example represents an enemy detachment) is correlated with object L (which for example is a series of chips or numbers). Finally, the rules for operations with the language objects and the relations among them (for example of the type $6 + 3 = 9$ and so on) correspond to associations between concepts in the brain. This concludes the analogy between models realized by means of language and models created by the neuron nets of the brain.

If the tool is a continuation of the human hand, then language is a continuation of the human brain. It serves the same purpose as the brain: to increase the vitality of the species by creating models of the environment. It continues the work of the brain using material lying outside the physical body, basing itself on models (concepts and associations) of the pre-language period which are realized in nerve nets. It is as if the human being had stepped across the boundary of his own brain. This transition, this establishment of a relationship between internal and external material, became possible owing to the capability for control of associating, which was expressed in the creation of language.

The two functions of language, communication and modeling, are inseparably interconnected. We gave counting on the fingers as an example of a model which arises only thanks to language and which cannot exist without language. When language is used for communication it performs a more modest task: it fixes a model which already exists in someone's brain. Phrases such as "It is raining" or "There are wolves in the neighboring forest," or more abstract ones such as "poisonous adder" or "fire extinguishes water," are models of reality. When one person communicates this to another the associations, which were formerly in the head of the first person only, become established in the head of the second.

Owing to the existence of language human society differs fundamentally from animal communities. In the animal world members of a community communicate only on the level of functions related to food and reproduction. Members of human society communicate not only on this level, but also on the highest level of their individual organization, on the level of modeling the external world by means of the association of representations. People *have contact by brain.* Language is not only a continuation of each individual brain but also a general, unitary continuation of the brains of all members of society. It is a collective model of reality on whose refinement all members of society are working, one that stores the experience of preceding generations.

■ SOCIAL INTEGRATION

THE METASYSTEM transition in the structure of the brain, control of associating, generated a new process, that of social integration—the unification of human individuals into a certain new type of whole unit: human society. All human history has gone forward under the banner of social integration; relations among people are growing qualitatively and quantitatively. This process is taking place at the present time, very intensively in fact, and no one can say for sure how far it will go.

Social integration is a metasystem transition; it leads to the appearance of a new level of organization of matter, the social sphere. Communities of animals can be viewed as the first (and unsuccessful) attempts to make this transition. We know communities of animals, for example ants, in which certain individuals are so adapted to life within the community that they cannot live outside it. The anthill may with full justification be called a single organism; that is how far interaction among individuals and specialization of them has gone in it. But this interaction remains at the level of the lowest functions. There is no "contact among brains." There is no creation of new models of reality. No fundamentally new possibilities are opened up because of the joining of ants into a society; they are frozen in their development. The anthill is, of course, a metasystem in relation to the individual ant. The integration of individuals takes place. However, this is not a new stage in evolution, but merely a digression, a blind alley. In Russian literature the word *sotsial'nyi* (social) which has the same literal meaning as the word *obshchestvennyi*, has traditionally been used to apply to human society only, thus emphasizing the fundamental difference between it and animal society. That is why I use the term *sotsial'nyi* here and it is how the phrases "social sphere" and "social integration" must be understood.

Thus, attempts by nature to form a new stage in the organization of matter by integrating multicellular organisms had no significant results for a long time: there was no appropriate material. A metasystem transition in the structure of the brain was needed in order for individuals to acquire the capability to make the necessary connections. One other consequence of the control of associating is very important for development of the social sphere. This is the capability of the human being to go beyond instinct, to construct plans of action that are completely unrelated to instinct and sometimes even contradict it. These two characteristics make the human being a *social being*—that is, material suitable for building human society, the social unit as opposed to the individual. The word "material" in reference to human beings sounds wrong, somehow degrading. Do you in fact think there is some kind of higher being who is building society using human beings as material? Of course not. The human being himself is the creator. And this is not some abstract Human (with a capital letter), but a concrete human, a human personality, an individual. Everything that society possesses has been produced by the creativity of human individuals. But at the same time (such is the dialectic of the

	1 Chemical foundations of life
Chemical era	
	2 Movement
	3 Irritability (simple reflex)
Cybernetic era	4 Nerve net (complex reflex)
	5 Associating (conditioned reflex)
	6 Thought
Era of reason	
	7 Social integration, culture

Figure 4.2. Stages in the evolution of life.

relationship between the personality and society) the human being is significant only to the extent that he or she is significant for society. This must not be understood, of course, to mean that someone who is not recognized is not a genius. A person may oppose the entire society, that is to say all those people living at a given moment, but at the same time be guided by the interests of society, the logic of society's development. There are two levels of the organization of matter: the animal level, for which the highest laws are the instincts of self-preservation and reproduction, and the human level, which means human society. Everything in the human being that we call distinctly human is a product of the development of society. The human being as a purely biological (pre-social) being is nothing but the possibility of the human being in the full sense of the word. If there is any logic at all in human actions it is either the logic of animal instincts or the logic of society's development (possibly veiled and not recognized as such). There is simply nowhere else to find any other logic. Therefore, although there is no being to whom the human, acting as creator, is subordinate, the human being is nevertheless subordinate to some highest law of evolution of the universe and, it may be said, is the material for the action of this law.

■ THE SUPER-BEING

THE APPEARANCE of human society is a large-scale metasystem transition in which the subsystems being integrated are whole organisms. On this level it may be compared with the development of multicellular organisms from unicellular ones. But its significance, its revolutionary importance, is immeasurably greater. And if it is to be compared to anything it can only be compared to the actual emergence of life. For the appearance of the human being signifies the appearance of a new mechanism for more complex organization of matter, a new mechanism for evolution of the universe. Before the human being the development and refinement of the highest level of organization, the brain device, occurred only as a result of the struggle for existence and natural selection. This was a slow process requiring the passage of many generations. In human society the de-

THE HUMAN BEING **97**

velopment of language and culture is a result of the creative efforts of all its members. The necessary selection of variants for increasing the complexity of organization of matter by trial and error now takes place in the human head. It can take place at the level of intuition—as the result of sudden enlightenment and inspiration—or it may break down into distinct, clearly recognized steps. But in one way or the other it becomes inseparable from the willed act of the human personality. This process differs fundamentally from the process of natural selection and takes place incomparably faster, but in both its function (constructing and using models of the environment) and in its results (growth in the total mass of living matter and its influence on nonliving matter) it is completely analogous to the earlier process and is its natural continuation. The human being becomes the point of concentration for Cosmic Creativity. The pace of evolution accelerates manyfold.

Society can be viewed as a single super-being. Its "body" is the body of all people plus the objects that have been and are being created by the people: clothing, dwellings, machines, books, and the like. Its "physiology" is the physiology of all people plus the *culture* of society—that is, a certain method of controlling the physical component of the social body and the way that people think. The emergence and development of human society marks the beginning of a new (the seventh in our count) stage in the evolution of life. The functional formula of the metasystem transition from the sixth stage to the seventh is:

$$\text{control of } thinking = culture$$

Language is the most important constituent part of culture. It fulfills the functions of a nervous system. As in the nervous system of a multicellular organism, its first function historically and logically is the communicative function, the exchange of information among subsystems and coordination of their activity. In the process of carrying out this function language, again precisely like the nervous system "one step lower," receives a second function: modeling the environment. And just as stages related to metasystem transitions can be identified in the development of the brain, so the development of language models takes place (as we shall see) by successive metasystem transitions in the structure of language.

The parallels between society and a multicellular organism have long been noted. But the question is: what are we to make of these parallels? It is possible to consider them, if not random, then at least superficial and insignificant—something like the resemblance between the boom of a hoisting crane and the human arm. But the cybernetic approach brings us to another point of view according to which the analogy between society and the organism has a profound meaning, testifying to the existence of extraordinarily general laws of evolution that exist at all levels of the organization of matter and pointing out to us the direction of society's development. This point of view conceals in itself a great danger that in vulgarized form it can easily lead to the conception of a fascist-type totalitarian state. In chapter 14, in our discussion of the problem of creative freedom of the personality, we shall consider this question in greater detail too. For now we shall note that the possibility a theory may be vulgarized is in no way an argument against its truth. The branch of modern science called cybernetics gives us concepts that describe the evolutionary process at both the level of intracellular structures and the level of social phenomena. The fundamental unity of the evolutionary process at all levels of organization is transformed from a philosophical view to a scientifically substantiated fact. When thinking of the destiny of humanity and its role in the universe one cannot ignore this fact.

To emphasize the cosmic importance of reason the French scientists Leroy and Teilhard de Chardin introduced the term "noosphere" (that is, the sphere of reason) to signify that part of the biosphere where reason reigns. These ideas were taken up by V. P. Vernadsky (see his article entitled "A Few Words About the Noosphere"). In the preface to his main work Le phénomène humain (The Phenomenon of Man, translated by B. Wall, New York: Harper and Row Torchbook ed., 1965, p 36) Teilhard de Chardin writes:

> In fact I doubt whether there is a more decisive moment for a thinking being than when the scales fall from his eyes and he discovers

that he is not an isolated unit lost in the cosmic solitudes, and realizes that a universal will to live converges and is hominized in him.

In such a vision man is seen not as a static center of the world—as he for long believed himself to be—but as the axis and leading shoot of evolution, which is something much finer.

CHAPTER FIVE
From Step to Step

■ MATERIAL AND SPIRITUAL CULTURE

A DISTINCTION IS MADE between "material" and "spiritual" culture. We have put these words in quotes (the first time; henceforth they will parade themselves in the customary way, without quotes) because the distinction between these two manifestations of culture is arbitrary and the terms themselves do not reflect this difference very well. Material culture is taken to include society's productive forces and everything linked with them, while spiritual culture includes art, religion, science, and philosophy. If we were to attempt to formulate the principle on the basis of which this distinction is made, the following would probably be the best way: material culture is called upon to satisfy those needs which are common to humans and animals (material needs), while spiritual culture satisfies needs which, we think, are specifically human (spiritual needs). Clearly this distinction does not coincide with the distinction between material and spiritual on the philosophical level.

The phenomenon of science, the chief subject of this book, is a part of spiritual culture. But science emerges at a comparatively late stage in the development of society and we cannot discuss this moment until we have covered all the preceding stages. Therefore we cannot bypass material culture without saying at least a few words about it. This is especially true because in the development of material culture we find one highly interesting effect which the metasystem transition sometimes yields.

■ THE STAIRWAY EFFECT

A baby is playing on the bottom step of a gigantic stone stairway. The steps are high and the baby cannot get to the next step. It wants very much to see what is going on there; now and again it tries to grab hold of the edge of the step and clamber up, but it cannot. . . . The years pass. The baby grows and then one fine day it is suddenly able to surmount this obstacle. It climbs up to the next step, which has so long attracted it, and sees that there is yet another step above it. The child is now able to climb it too and thus, mounting step after step, the child goes higher and higher. As long as the child was unable to get from one step to the next it could not ascend even a centimeter; but as soon as it learned how, not only the next step but the entire stairway became accessible.

A schematic representation of this "stairway effect" is shown in figure 5.1. The stairway effect forms the basis of many instances in which small quantitative transitions become large qualitative ones. Let us take as an example the classical illustration of Hegel's law of the change of quantity into quality: the crystallization of a liquid when the temperature drops below its melting point. The ability of a molecule oscillating near a certain equilibrium position to hold several adjacent molecules near their equilibrium positions is precisely the "capability of transition to the next step." When this capability manifests itself as a result of a drop in temperature (decrease in the

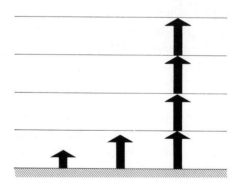

Figure 5.1. The stairway effect.

amplitude of oscillations) the process of crystallization begins and "step by step" the positions of the molecules are set in order. Another well-known example is the chain reaction. In this case the transition to the next step is the self-reproduction of the reagents as a result of the reaction. In physical systems where all relationships important for the behavior of the system as a whole are statistical in nature, the stairway effect also manifests itself statistically; the criterion of the possibility of transition to the next step is quantitative and statistical. In this case the stairway effect can be equated with the chain reaction, if the latter term is understood in the very broadest sense.

■ THE SCALE OF THE METASYSTEM TRANSITION

WE ARE MORE INTERESTED in the case where the transition to the next step is qualitative, specifically the metasystem transition. For the stairway effect to occur in this case it is clearly necessary for system X, which is undergoing the metasystem transition (see figure 5.2), to itself remain a subsystem of some broader system Y, within which conditions are secured and maintained for multiple transitions "from step to step"—the metasystem transition beyond subsystem X. We shall call such a system Y an *ultrametasystem* in relation to the series X. X', X'', . . . and so on. Let us take a more detailed look at the question of the connection between the metasystem transition and the system–subsystem relation.

We have already encountered metasystem transitions of different

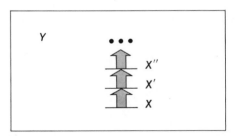

Figure 5.2. The stairway effect within ultrametasystem Y. The arrows indicate changes taking place in time. ·

scale. Metasystem transitions in the structure of the brain are carried out within the organism and do not involve the entire organism. Social integration is a metasystem transition in relation to the organism as a whole, but it does not take humanity outside of the biogeographic community, the system of interacting living beings on a world scale. There is always a system Y which includes the given system X as its subsystem. The only possible exception is the universe as a whole, the system Z which by definition is not part of any other system. We say "possible" exception because we do not know whether the universe can be considered a system in the same sense as known, finite systems.

Now let us look in the opposite direction, from the large to the small, from the whole to the part. What happens in system X when it evolves without undergoing a metasystem transition? Suppose that a certain subsystem W of system X makes a metasystem transition (see figure 5.3 below). This means that system W is replaced by system W', which in relation to W is a metasystem (and contains a whole series of W-type subsystems) but in relation to X is a subsystem analogous to W and performs the same functions in X as W had been performing, only probably better. Depending on the role of subsystem W in system X, the replacement of W with W' will be more or less important for X. In reviewing the stages in the evolution of living beings during the cybernetic period we substituted the organism as a whole for X and the highest stage of control of the organism for W. There-

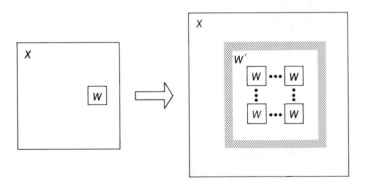

Figure 5.3. The metasystem transition $W \rightarrow W'$ within system X.

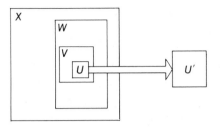

Figure 5.4. Metasystem transition at one of the lower levels of organization.

fore the metasystem transition $W \rightarrow W'$ was of paramount importance for X. But a metasystem transition may also occur somewhere "in the provinces," at one of the lower levels of organization (see figure 5.4). Suppose W is one of the subsystems of X, V is one of the subsystems of W, and U is one of the subsystems of V. The metasystem transition $U \rightarrow U'$ may greatly improve the functioning of V, and consequently the functioning of W also, although to a lesser degree, and finally, to an even smaller degree, the functioning of X. Thus, evolutionary changes in X, even though they are not very significant, may be caused by a metasystem transition at just one of the lower levels of the structure.

These observations provide new material for assessing quantitative and qualitative changes in the process of development. If system X contains homogeneous subsystems W and the number of these subsystems increases we call such a change *quantitative*. We shall unquestionably classify the metasystem transition as a *qualitative* change. We can assume that any qualitative change is caused by a metasystem transition at some particular level of the structure of the system. Considering the mechanics of evolution described above (replication of systems plus the trial and error method) this assumption is highly probable.

■ TOOLS FOR PRODUCING TOOLS

LET US RETURN to material culture and the stairway effect. The objects and implements of labor are parts, subsystems of the system we have called the "super-being," which emerges with the development of human society. Now we shall simply call this super-being culture, meaning by this both its physical "body" and its method of functioning ("physiology"), depending on the context. Therefore, the objects and implements of labor are subsystems of culture. They may possess their own complex structure and, depending upon how they are used, they may be part of larger subsystems of culture which also have their own internal structures.

Specifically, the division of material subsystems into objects of labor and implements of labor (tools) is in itself profoundly meaningful and reflects the structure of production. When a human being applies tool B to objects of a certain class A, this tool, together with objects A, forms a metasystem in relation to subsystems A. Indeed, subsystem B acts directly upon sybsystems A and is specially created for this purpose. (Of course, this action does not take place without the participation of the human hand and mind, which are part of any production system.) Thus, the appearance of a tool for working on certain objects that had not previously been worked on is a *metasystem transition* within the production system. As we have seen, the ability to create tools is one of the first results of the development of human traits; and because the human being remains the permanent moving force of the production system, the metasystem transition from object of labor to implement of labor may be repeated as many times as one likes. After having created tool B to work on objects in class A the human being begins to think of ways to improve the tool and manufactures tool C to use in making tools of class B. He does not stop here; he makes tool D to improve tools of class C, and so on. The implement of labor invariably becomes an object of labor. This is the stairway effect. It is important to assimilate the very principle of making tools (learning to climb up a step). After this assimilation everything follows of its own accord: the production system becomes an ultrametasystem capable of development. The result of this process is modern industry, a highly complex multilevel system which uses natural materials and step by step converts them into its "body"—structures, machines, and instruments—just as the living organism digests the food it has eaten.

■ THE LOWER PALEOLITHIC

LET US CONSTRUCT a general outline of the development of material culture. The history of culture before the emergence of metalworking is divided into two ages: the Paleolithic (Old Stone Age) and the Neolithic (New Stone Age). In each age distinct cultures are identified, which differ by geographic region and the time of their existence. The cultures which have been found by archaeological excavation have been given names derived from the names of the places where they were first discovered.

Traces of Paleolithic culture have been found in many regions of Europe, Asia, and Africa. They enable us to confidently make a periodization of the development of culture in the Paleolithic and divide the age into a number of stages (epochs) which are universally important for all geographic regions.

The most ancient stages are the so-called Chellean, followed by the Acheulean and then the Mousterian. These three stages are joined together under the common name Lower (or Early) Paleolithic. The beginning of the Lower Paleolithic is dated about 700,000 years ago and the end (the late Mousterian culture) is dated about 40,000 years ago.

The Chellean and Acheulean cultures know just one type of stone tool—the hand ax. The Chellean hand ax is very primitive; it is a stone crudely flaked on two sides, resembling a modern axhead in shape and size. The typical Acheulean hand ax is smaller and much better made; it has carefully sharpened edges. In addition, signs of the use of fire are found at Acheulean sites.

The tools of the Mousterian culture reveal a clear differentiation. Here we distinguish at least two unquestionably distinct types of stone tools: points and scrapers. Stoneworking technique is considerably higher in the Mousterian period than in the Acheulean. Objects made of bone and horn appear. Fire is universally used. We do not know whether the Mousterians were able to make fire, but it is clear that they were able to preserve it.

In a biological sense the human being of the Lower Paleolithic was not yet the modern form. The Chellean and Acheulean cultures belonged to people (or semipeople?) of the *Pithecanthropus* and *Sinanthropus* types. The Mousterian was the culture of the Neanderthals. In the Lower Paleolithic the development of techniques for making tools (not only from stone but also from wood and other materials which have not survived until our day) proceeded parallel with the development of human physical and mental capabilities, with human evolution as a species. The increase in brain size is the most convincing evidence of this evolution. The following table shows the capacity of the cranial cavity in fossil forms of man, the anthropoid apes, and modern man:

Gorilla	600–685 cm³
Pithecanthropus	800–900 cm³
Sinanthropus	1,000–1,100 cm³
Neanderthal	1,100–1,600 cm³
Modern Man	1,200–1,700 cm³.

Let us note that although the Neanderthal brain is only slightly smaller in volume than the brain of modern man it has significantly smaller frontal lobes and they play the chief role in thinking. The frontal lobes of the brain appear to be the principal storage area for "arbitrary" associations.

■ THE UPPER PALEOLITHIC

AT THE BOUNDARY between the Lower and Upper Paleolithic (approximately 40,000 years ago) the process of establishment of the human being concludes. The human being of the Upper Paleolithic is, in biological terms, modern man: *Homo sapiens*. From this moment onward nature invests all its "evolutionary energy" in the culture of human society, not in the biology of the human individual.

Three cultures are distinguished in the Upper Paleolithic: Aurignacian, Solutrean, and Magdalenian. The first two are very close and are joined together in a single cultural epoch: the Aurignac-Solutrean. The beginning of this epoch is coincident with the end of the Mousterian epoch. Several sites have been found containing bones of both Neanderthals and modern man. It follows from this that the last evolutionary change, which completed the formation of modern man, was very significant and the new people quickly supplanted the Neanderthals.

In the Aurignac-Solutrean epoch, stone-working technique made great advances in comparison with the Mousterian epoch. Various types of tools and weapons can be found: blades, spears, javelins, chisels, scrapers, and awls. Bone and horn were used extensively. Sewing appeared, as evidenced by needles which have been found. In one of the monuments of Solutrean culture a case made of bird bone and containing a whole assortment of bone needles was found, as was a bone fishing hook. By the Magdalenian epoch (about 15,000 years ago) throwing spears and harpoons had appeared. A noteworthy difference between the Upper Paleolithic and the Lower is the emergence of visual art. Cave drawing appeared in the Aurignac-Solutrean epoch and reached its peak in the Magdalenian. Many pictures (primarily of animals) have been found whose expressiveness, brevity, and exactness in conveying nature amaze even the modern viewer. Sculptured images and objects used for ornamentation also appear. There are two points of view on the question of the origin of art: the first claims art is derived from magic rituals, the second from esthetic and cognitive goals. However, when we consider the nature of primitive thinking (as we shall below) the difference between these two sources is insignificant.

Looking at material production as a system, the crucial difference between the Upper Paleolithic and the Lower is the appearance of composite implements (for example, a spear with a stone point). Their appearance can be viewed as a metasystem transition, because in making a composite implement a system is created from subsystems. Before, the maker would have viewed the two components as independent entities (the point as a piercing stone tool and the pole as a stick or wooden spear). This is not a simple transition; even in historical times, there could be found a group of people (the indigenous inhabitants of the island of Tasmania) who did not know composite implements.

The Tasmanians no longer exist as an ethnic group. The last pure-blooded Tasmanian woman died in 1877. The information about the Tasmanian culture that has been preserved is inadequate and sometimes contradictory. Nonetheless, they may certainly be considered the most backward human group of all those known by ethnography. Their isolation from the rest of the human race (the Tasmanians' nearest neighbors, the Australian aborigines, were almost equally backward) and the impoverished nature of the island, in particular the absence of animals larger than the kangaroo, played parts in this. With due regard for differences in natural conditions, the culture of the Tasmanians may be compared to the Aurignac-Solutrean culture in its early stages. The Tasmanians had the stone hand ax, sharp point, crudely shaped stone cutting tool, wooden club (two types, for hand use and throwing), wooden spear, stick for digging up edible roots, and wooden spade for scraping mussels off rocks. In addition they were able to weave string and sacks (baskets) from grass or hair. They made fire by friction. But, to again repeat, they were not able to make composite tools—for example, to attach a stone working part to a wooden handle.

■ THE NEOLITHIC REVOLUTION

UNLIKE THE PALEOLITHIC CULTURES, the Neolithic cultures (which are known from both archaeological and ethnographic findings) show great diversity, specificity, and local characteristics. In terms of techniques of producing tools the Neolithic is an elaboration of the qualitative jump (metasystem transition) made in the late Paleolithic: composite tools made using other tools. Following this route human beings made a series of outstanding advances, the most remarkable of which is clearly the invention of the bow. Great changes also took place in clothing and in the construction of dwellings.

Although the Neolithic cannot boast of a large-scale metasystem transition in regard to tool manufacture, a metasystem transition of enormous importance nevertheless did occur during this period. It concerned the overall method of obtaining food (and therefore it indirectly involved tools also). This was the transition from hunting and gathering to livestock herding and farming—sometimes called the Neolithic revolution. The animal and plant worlds, which until this had been only external, uncontrolled sources of food, now became subject to active influence and control by human beings. The effects of this transition spread steadily. We are thus dealing with a typical metasystem transition.

Archaeologists date the emergence of farming and livestock herding to about 7,000 years ago, emphasizing that this is an approximate date. The most ancient cereal crops were wheat, millet, barley, and rice. Rye and oats appeared later. The first domesticated animal was the dog. Its domestication is dated in the Early Neolithic, before the emergence of farming. With the transition to farming, people domesticated the pig, sheep, goat, and cow. Later, during the age of metal, the domesticated horse and camel appeared.

■ THE AGE OF METAL

THE AGE OF METAL is the next page in the history of human culture after the Neolithic. The transition to melting metal marks a metasystem transition in the system of production. Whereas the material used earlier to make tools (wood, stone, bone, and the like) was something given and ready to use, now a process, melting metal, emerged and it was directed not to making a tool but rather to making the material for the tool. As a result people received new materials with needed characteristics that were not found in nature. First there was bronze, then iron, various grades of steel, glass, paper, and rubber. From the point of view of the structure of production the age of metal should be called the *age of materials*. Strictly speaking, such crafts as leather tanning and pottery, which orginated earlier than metal production, should be viewed as the beginning of the metasystem transition to the age of materials. But there is a crucial phase in each metasystem transition when the advantages of creating the new level in the system become obvious and indisputable. For the age of materials this phase was the production of metals, especially iron.

The most ancient traces of bronze in Mesopotamia and Egypt date to the 4th millennium B.C. Iron ore began to be melted by 1300 B.C.

■ THE INDUSTRIAL REVOLUTIONS

THE NEXT qualitative jump in the system of production was the use of sources of energy other than the muscular energy of human beings and animals. This, of course, is also a metasystem transition because a new level of the system emerges: the level of engines which control the movement of the working parts of the machine. The first industrial revolution (eighteenth century) radically changed the entire appearance of production. Improvement of engines becomes the leitmotif of technical progress. First there was the steam engine, then the internal combustion engine, and then the electric motor. The age of materials was followed by the age of energy.

Finally, our day is witness to one more metasystem transition in the structure of production. A new level is emerging, the level of control of engines. The second industrial revolution is beginning, and it is obvious that it will have a greater effect on the overall makeup of the system of production than even the first did. The age of energy is being replaced by the age of information. Automation of production processes and the introduction of computers into national economies lead to growth in labor productivity which is even more rapid than before and give the production system the character of an autonomous, self-controlling system.

■ THE QUANTUM OF DEVELOPMENT

THE SIMILARITY between successive stages in the development of technology and the functions of biological objects has long been noted. The production of industrial materials can be correlated with the formation and growth of living tissue. The use of engines corresponds to the work of muscles, and automatic control and transmission of information corresponds to the functioning of the nervous system. This parallel exists despite the fundamental difference in the nature of biological and technical systems and the completely different factors that cause their development. Nonetheless, the similarity in the stages of development is far from accidental. It arises because all processes of development have one common feature: development always takes place by successive metasystem transitions. The metasystem transition is, if you like, the elementary unit, the universal quantum of development. Therefore it is not surprising in the least that having compared the initial stage of development of two different systems—for example industrial materials and living tissue—we receive a natural correlation among later stages, which are formed by the accumulation of these universal quanta.

■ THE EVOLUTION OF THOUGHT

OUR NEXT TASK on the historical plane is to analyze the development of thought beginning with the most ancient phase about which we have reliable information. This phase is primitive society with Late Paleolithic and Early Neolithic culture. But before speaking of primitive thinking, before "putting ourselves in the role" of primitive people, we shall investigate thinking in general, using both the modern thinking apparatus as an investigative tool and modern thinking as an object of investigation that is directly accessible to each of us from personal experience. This is essential in order that we may clearly see the difference between primitive thinking and modern thinking and the general direction of the development of thinking. The investigation we are preparing to undertake in the next two chapters can be defined as a cybernetic approach to the basic concepts of logic and to the problem of the relationship between language and thinking.

CHAPTER SIX
Logical Analysis of Language

■ ABOUT CONCEPTS AGAIN

LET US BEGIN with the most fundamental concept of logic, the concept of the "concept." In chapter 2 we gave a cybernetic definition of the concept in its Aristotelian version—as a set of situations at the input of a cybernetic system. To master a concept means to be able to recognize it, that is, to be able to determine whether or not any given situation belongs to the set that characterizes this concept. This definition applies equally to complex cybernetic systems of natural origin about whose organization we have only a general idea (for example, the brain of an animal) and to those relatively simple systems we ourselves create for applied and research purposes.

In the first case we arrive at the conclusion, that the system recognizes a certain concept, by observing external manifestations of the system's activity. For example, when we see that a dog becomes happily excited when it hears its master's voice and responds in a completely different way to all other sounds we conclude that the dog has the concept of "master's voice." This concept develops in the dog naturally, without any special effort by the experimenter. To determine the maximum capabilities of the dog brain the experimenter may create unusual conditions for the animal and watch its reaction. I. P. Pavlov and his school conducted many such experiments. If a dog is shown plywood circles and squares of different sizes and colors and is fed after the presentation of a circle and punished after the presentation of a square, the dog will learn to distinguish the circle and the square and will respond differently when these shapes are presented. Thus, the dog is capable of recognizing certain general (abstract) concepts—in this case the concepts of circle and square abstracted from the features of size and color. This means we must conclude that the dog possesses the abstract concepts of "circle" and "square."

But no sooner do we say this than we begin to feel that perhaps this conclusion was too hasty. The statement that the dog can possess the concept of "master's voice" (referring, of course, to the voice of a specific person) can be accepted without reservation, but the statement that the concept of square is accessible to the dog seems true in one sense and not in another. We shall take note of this now and return to the question later. In the meantime let us examine the dog's mental capabilities by indicating the very simple concepts known to be inaccessible to the animal. Suppose that you show the dog a box divided into two parts, each of which contains several billard balls. You want to force the dog to distinguish the case where the number of balls in each part is equal from the case where the number of balls in the parts differs. It is a safe bet that no matter how much you feed the dog and no matter how much you beat it you will not achieve your purpose. The concept of different numbers is inaccessible to the dog.

Cybernetic systems possessing the ability to recognize concepts are also created artificially. Their importance is steadily growing in connection with cybernetic science and production. The development of artificial recognition devices (discriminators) plays a crucial part in understanding the general principles and concrete mechanisms of the working of the brain. These devices serve as models with which people try to lift the veil from the process of thinking. The creation of an "artificial brain" which performs, at least partially, the same functions as the natural brain provides indications of how to approach investigation of the activity of the natural brain. It is interesting that one of the first results of comparing artificial and natural recognition systems was the conclusion that natural systems are very narrowly goal-directed and specialized. Within their own specialization they reach a high level of refinement, but they are completely impotent when the problem goes beyond this framework. Recognizing a person by voice is an extremely difficult problem for artificial cybernetic devices, but the brain of a dog resolves it easily. At the same time the problem of comparing the number of billard balls, which is very simple for an artificial system, is beyond the the ability of a dog.

In chapter 2 we considered a cybernetic discriminator that was fed information by signals from light-sensitive receptors arranged on a screen. We called the situation, that is to say the aggregate of values of all signals from the receptors, the "picture"; it coincides with the image on the screen with a precision down to semitones. This device (picture discriminator) will serve as an illustration in this chapter too.

■ ATTRIBUTES AND RELATIONS

SO FAR, the examples of concepts we have given have fit within the definition of the concept as a set of situations. But as it turns out, this definition does not apply to every concept that seems intuitively clear to us and manifests itself in language. For example, let us take the concept expressed by the prepositions "inside" or "in." For those who do not like to see a concept expressed by a preposition, we can express it by phrases such as "to be located inside" or "location in." This concept is applicable to a device to whose input "pictures" are fed. In figure 6.1 for example, spot A is *inside* contour B. But can we correlate the concept of "inside" to some definite set of pictures? No, we cannot. This can be seen, for example, from consideration of

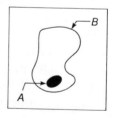

Figure 6.1.

116   ANALYSIS OF LANGUAGE

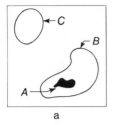 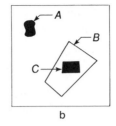

a   b

Figure 6.2.

the picture shown in figure 6.2. In figure 6.2a spot A is inside contour B, but not inside contour C. In figure 6.2b, spot A is outside contour B while spot C is inside it.

Could these pictures be classed with a set of situations for "inside" which we would have to construct? Any answer will be unsatisfactory and arbitrary, because the question itself is meaningless. The concept of "inside" does not characterize a picture (situation) as a whole but rather the relation between two definite objects, details in the picture. As long as these objects are not indicated, a definite spot and a definite contour, it is meaningless to ask the question "inside or not"?

■ ARISTOTELIAN LOGIC

WE SHALL call the concepts that express attributes of the situation as a whole "Aristotelian," because Aristotle's logic is simply a consistent theory of the correct use of such concepts. For each Aristotelian concept there is a definite corresponding set of situations, specifically those situations in which the attribute expressed by this concept occurs. Therefore the Aristotelian concept can also be described as a certain set or class of situations (phenomena, objects in that extremely general sense in which these terms are used here; they are all equivalent to one another and to the term "something" [in Russian *nechto*], which is the most precise but also the most inconvenient because of the difficulties with Russian grammar its use entails). Therefore all the laws of Aristotelian logic can also easily be derived from the simplest properties of operations on sets.

For example, let us take the classical syllogism:

All men are mortal
Socrates is a man
Therefore Socrates is mortal.

Three Aristotelian concepts participate in this reasoning: "man," "mortal," and "Socrates." The concept of "man" is characterized by the set of situations in which we say, "This is a man." The same thing applies to the other concepts. To make the properties of the sets graphically clear let us represent each situation as a point within a certain square (see figure 6.3). Then this square will embody the set of all conceivable situations corresponding to the maximally general concept "something." The other concepts, to which different sets of points correspond, will be shown by different areas inside the square. The statement that "all men are mortal," in other words "every man is mortal," signifies that every point included in the area of "man" is also within the area of "mortal" ("mortal being"), which is to say that the "man" area is entirely inside the "mortal" area. In exactly the same way the second premise of the syllogism means that the "Socrates" area is entirely inside the "man" area. It follows from

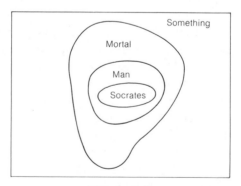

Figure 6.3.

118 ANALYSIS OF LANGUAGE

this that the "Socrates" area is within the "mortal" area, or in other words the statement "Socrates is mortal" is true.

Figure 6.4 demonstrates the correctness of the following deduction rule ("disamis" in logical terminology):

All A are B
Some A are C
Therefore, some B are C.

Aristotle's logic played an important role in the development of European culture. But it does not go deeply enough into the structure of our thinking; it is not able to reflect the process of breaking situations up into distinct parts (objects) and investigating the relations among these parts. In discussing the attributes of objects Aristotelian logic is completely adequate, because an isolated object can be pictured as a certain situation. Forming the set of such situation-objects, we obtain an abstract concept that expresses one of the properties of the object. Things are different with relations. Aristotelian logic can express the concept of the aggregate of objects which are in a given relation, but it has no means for expressing the concept of the relation as such. We can represent a set of pictures that have the form of contours with spots in the middle; this set generates the Aristotelian concept (attribute) of "being a contour with a spot in the middle." But there is no Aristotelian concept for "being inside." Aristotle's logic is too global and superficial.

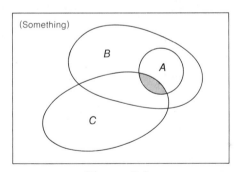

Figure 6.4.

Let us take the following deduction:

<u>Ivan is Peter's brother</u>
Therefore, Peter is Ivan's brother.

The inference is correct, but to substantiate it one must state openly the premise which is taken for granted here: that the relation of "brothers" is symmetrical. This premise can be expressed by the phrase: "If x is the brother of y, then y is the brother of x."

In this the letters x and y represent any persons of the male sex. But this symbolism goes beyond the limits of Aristotelian logic.

Can this syllogism be expressed in the language of Aristotle's logic? It can if we consider not individual people but pairs of people—or, more exactly, ordered pairs, which is to say pairs where one person is assigned the number one and the other receives the number two. Here is this syllogism, which is completely proper from the point of view of Aristotelian logic.

The pair (Ivan and Peter) possesses the attribute: the first is the brother of the second

Each pair possessing the attribute: "the first is the brother of the second" possesses the attribute: "the second is the brother of the first"

_____

Therefore the pair (Ivan and Peter) has the attribute: "the second is the brother of the first."

Even though formally it is the same reasoning used before, this clumsy syllogism misses the mark because it does not reflect the main element in our initial syllogism, the symmetry in the relation of "brothers." The attributes "the first is the brother of the second" and "the second is the brother of the first" are in no way broken down, in no way connected with one another, and in no way connected with the fact that they are applied to objects which have the appearance of an ordered pair.

It was not accidental that we began our cybernetic investigation of concepts from Aristotelian concepts. They are simpler because they permit definition exclusively in terms of input and output states without referring to the internal structure of the recognition system.

The same thing occurred in the history of human thought. People first became aware of the existence of Aristotelian concepts; awareness of relations came only much later.

Because the chief thing in mathematics is to investigate relations among objects, Aristotelian logic is completely inadequate for expressing mathematical proofs. This was noted long ago; the examples from mathematics which traditional logic uses speak for themselves: they are extremely primitive and uninteresting. Until the very end of the nineteenth century, when a new ("mathematical") logic began to be created, mathematics and logic developed independently of one another.

■ HEGEL'S DIALECTIC

IN PHILOSOPHY Hegel delivered the decisive blow against Aristotelian logic. With his dialectic he showed that the world must be viewed not as an aggregate of objects that possess certain attributes, but rather as an aggregate of objects that stand in certain relations to one another. This does not exclude attributes from consideration, of course, for the concept of the relation is much broader than the concept of the attribute.

A relation may be defined for any number of objects. Specifically, the number of objects may be one; such a relation is an attribute, a property. Pair relations, that is to say relations between two objects, are the clearest intuitively and at the same time the most important. Two is the minimum number of objects for which the relation ceases to be an attribute, and becomes a relation proper. The number two lies at the foundation of the Hegelian method, which is reflected in the very term "dialectic."

The most important features of Hegel's dialectic follow directly from the description of phenomena in terms of relations, not attributes. Above all, what follows from this approach is the theory of the interaction and interrelatedness of everything that exists. Further: If two elements are in correspondence and do not contradict one another, they act as something whole and their common attributes become paramount while the interaction, the relation, between them withdraws to a secondary position. Relations among elements, objects, manifest themselves to the extent that they are relations of opposition, contradiction, and antagonism. Thus, the idea of the struggle of opposites plays an important part in Hegel.

When considering the relation between the state of an object at a given moment and the state of this same object at some other moment in time we come to the concept of change. Change is the relation between objects separated by a time interval. In the language that operates with attributes but not with relations, change cannot be expressed. The most that such language is capable of is depicting a series of states of the object which are in no way interconnected.

Zeno's aporia concerning the arrow in flight is a brilliant expression of this inability. Let us consider the arrow in flight. Take a certain moment in time. At this moment the arrow occupies a definite position in space. Take another moment. The arrow again occupies a completely determinate position in space. The same thing is true for any other moment. This means that the arrow always occupies a definite position in space. This means that it is standing in place. In Aristotelian concepts the world is represented as something static, frozen, or at best mechanically duplicated with certain variations. On the other hand, having made the investigation of relations its object, the dialectic studies things from the point of view of their change, movement, and development. It discloses the historical causality and relativity (from the word "relation"!) of things which are represented as unconditional and external when described in Aristotelian concepts. Combining the concept of opposition with the concept of the relation among states at successive moments in time generates the concept of the negation and the concept of the negation of the negation. The dialectic is dynamic and revolutionary.

In relation to Aristotelian logic, Hegel's dialectic acted as a destructive force—and not just because of its "general" revolutionary nature but also because it pointed out the many contradictions that arise when a description of phenomena which demands the language of relations is squeezed into the narrow framework of the language of attributes. In Hegel and his followers these contradictions were often surrounded with a certain exalted aura and, one might say, a semimystical significance. This reflected, on the one hand, the idealistic orientation of Hegel's philosophy, and on the other hand a general characteristic of new doctrines, theories, and movements: in the initial stages of their development, trying to liberate themselves from the old intellectual framework, they prefer a paradoxical, exaggerated form and become heroic and romantic. Hegel's dialectic is the heroic epoch of the new logic, when the old logical formalism had been broken but the new one was not yet created. Therefore things seemed contradictory and not subject to formalization ("dialectical") which later proved to be beautifully ordered and formalized. To modern thinking, which makes free use of the language of relations and is armed with analysis of logical concepts and constructions, the Hegelian style of thinking appears as obscure philosophizing about things which are clear. The following reasoning is a crudely simplified, caricature-like sketch of the Hegelian dialectical contradiction, showing the source from which this contradiction arises.

"Let us put the question: is the number 1,000 large or small? It is large because it is much more than one. It is small because it is much less than 1 million. This means that it is both large and small at the same time. A dialectical contradiction. What is large is at the same time not large, A is not-A."

The concepts of "large" and "small" were considered here as attributes of objects (numbers). In fact these are not attributes but concealed (by means of the grammatical category of the adjective) relations. An exact meaning can be given only to the concepts "larger" and "smaller." If we analyze the reasoning given above from this point of view it will prove to be simply nonsense. This caricature was not directed against Hegel (the credit due him for creating the new logic is indisputable) but rather against those who take an uncritical attitude toward Hegel's dialectical method and, in the second half of the twentieth century, propagate the style of thinking of the first half of the nineteenth century, ignoring the enormous progress made by logic in this century and a half.

■ MATHEMATICAL LOGIC

THE DECISIVE FACTOR in the advance of logic was the development of mathematical logic in the late nineteenth and early twentieth centuries. This process was generated by the needs of mathematics and was carried out by mathematicians. The gap between mathematics and logic was finally overcome. Having expanded its language and made it mathematical, logic became suitable for describing and investigating mathematical proof. On the other hand, mathematical methods began to be used to solve logical problems.

Having gained a base of operations in the field of mathematics, the new logic began to penetrate the natural sciences and philosophy. In this process the role of the mathematical element proper (the use of mathematical methods) declined. Nonetheless all modern logic is often called "mathematical" because of its language and origin.

■ OBJECTS AND STATEMENTS

BEFORE GOING on in our analysis of language and thinking we need to give a short sketch of modern logic and those concepts which are related to language. For now we will leave the concepts related to the logical deduction (proof) aside.

Modern logic divides everything that exists into *objects* and *statements*. In natural language statements are represented by sentences or groups of sentences and objects are depicted by words or combinations of words which make up the sentences. Examples of objects are "heron," "Uncle Kolya," or "kolkhoz chairman." Examples of statements are: "The heron died," or "Uncle Kolya was elected chairman of the kolkhoz." Objects are most often expressed by nouns, but this is not mandatory. For example, in the statement "To smoke is harmful," "to smoke" is the object. In application to mathematics objects are usually called terms and statements are called relations. Examples of terms are: (1) 3.14; (2) $ax^2 + bx + c$; (3) $\int_a^b f(z)dz$. Examples of relations are: (1) $ax^2 + bx + c = 0$; (2) $0 < z \leqslant 1$; (3) no matter what natural number $n > 1$ may be, a simple number p will be found which is a divisor of number n; (4) the sum of the squares of the legs of a right triangle is equal to the square of the hypotenuse.

In logic the concepts "object" and "statement" are considered primary, intuitively clear, and indefinable. The formal difference between them is that a statement may be said to be true or false. Thus, the examples (3) and (4) of mathematical relations above are true, while the first and second may be true or false depending on the values of the variables x and z. The concepts of truth and falsehood are not applicable to objects.

In logic objects and statements, which are considered elementary—meaning that they cannot be broken down into distinct constituent parts—are represented by letters. Objects are usually represented by small letters and statements by capital letters. We shall follow this system but we shall introduce one more convention. For clarity in writing and to reduce the number of verbal explanations we shall sometimes designate elementary objects and statements with words and phrases within quotation marks. Therefore phrases in quotes will be considered equal to letters.

Objects and statements which are not elementary are obviously constructed from other objects and statements. We must now point out the methods of construction. Where there are two types of elements (objects and statements) and assuming that the elements serving as building material all belong to one type, we find that there are four possible types of constructions. We have reduced them to the following table.

What Is Constructed	What It Is Constructed From	Name of the Construction
Statement	Statements	Logical Connective
Statement	Objects	Predicate
Object	Statements	—
Object	Objects	Function

■ LOGICAL CONNECTIVES

THERE ARE FIVE widely used logical connectives. Negation (depicted by the symbol $-$), conjunction (sign &), disjunction (sign V), material implication (sign $\supset$), and equivalence (sign $\equiv$).

The statement $-A$ (read "not A") means that statement A is false. In other words, $-A$ is true when A is false and it is false when A is true.

The statement A & B (read "A and B") signifies the assertion that both A and B are true. It is true only if both statement A and statement B are true.

The statement $A \lor B$ ("A or B") is true if at least one of the two statements A and B is true.

The statement $A \supset B$ is read "A entails B" or "if A then B." This is untrue if A is true and B is false but is considered true in all other cases.

Finally, the statement $A \equiv B$ is true if statements A and B are either both true or both false.

Parentheses are used to designate the structure of connections, similar to the way they are used in algebra to designate the order of performance of arithmetic operations. For example, the statement $-A \& B$ means "S is untrue but B is true," while the statement $- (A \& B)$ means "it is untrue that both A and B are true." And, just as in algebra, an order of seniority among connectives by the tightness of the bond is established to reduce the number of parentheses. Above we listed the connectives in order of decreasing tightness. For example, the conjunction is a tighter bond than implication and therefore the statement $A \supset B \& C$ is understood as $A \supset (B \& C)$, not as $(A \supset B) \& C$. This corresponds to algebra where $A + B \times C$ is the same as $A + (B \times C)$, but not the same as $(A + B) \times C$.

Let us give a few examples of composite statements. A common Russian tongue-twister is "The heron withered, the heron dried, the heron died" [in Russian, "tsaplya chakhla, tsaplya sokhla, tsaplya sdokhla"]. This statement may be written as follows: "The heron withered" & "the heron dried" & "the heron died." The relation $0 < Z < 1$ is the conjunction "$Z > 0$" & "$Z < 1$," while the relation $|Z| > 1$ is the disjunction "$Z > 1$" $\lor$ "$Z < -1$." The definition given above for the logical connective $\equiv$ can be written as follows:

$$[(A \equiv B) \supset (A \& B) \lor (-A \& -B)] \, \& \, [(A \& B) \lor (-A \& -B) \supset (A \equiv B)]$$

We will let the reader translate the following statement into conventional language: "The light is turned on" & "the bulb is not burning" $\supset$ "there is no electricity" $\lor$ "the plugs have burned out" $\lor$ "the bulb is burned out."

If we consider that statements can only be true or false, and consider nothing else about them, then the connectives we have listed are enough to express all conceivable constructions made of statements. Even two connectives are adequate—for example, negation and conjunction or negation and disjunction. This situation obtains, in particular, in relation to mathematical statements. Therefore other connectives are not used in mathematical logic.

But natural language reflects a greater diversity in the evaluation of statements than simply separating them into true statements and false. For example, a statement may be considered meaningless or implausible even though it is possible ("There are probably wolves in this forest"). Special branches of logic which introduce other connectives are devoted to these matters. For modern science (unlike classical mathematical logic) these branches are not very important and we shall not deal with them.

■ PREDICATES

A CONSTRUCTION that associates a statement with certain objects is called a predicate. Predicates are divided into one-place, two-place, three-place, and so on according to the number of objects they require. Functional notation is used to represent them. The predicate can be written as a function with unfilled places for variables, for example:

$$P(\ ),$$
$$L(\ ,\ ),$$
$$I(\ ,\ ,\ )$$

or in the form

$$P(x),$$
$$L(x,\ y),$$
$$I(x,\ y,\ z)$$

having stipulated that x, y, and z are object variables, that is, symbols which must in the last analysis be replaced by objects—although which objects is not yet known. But the second form of notation, strictly speaking, no longer represents a predicate; rather, it is a statement containing object variables. In addition to capital letters we shall also use words and phrases within quotation marks, for example: "red" (x) or "between" (x, y, x) and special mathematical signs such as $< (x, y)$.

The one-place predicate expresses an attribute of an object, while a predicate with more than one variable expresses a relation among objects. If the places for variables in the predicate are filled, then we are dealing with a statement which asserts the existence of the given attribute or relation. The statement "red" ("ball") means that the "ball" possesses the attribute "red." The construction $<(a, b)$ is equivalent to the relation (inequality) $a < b$.

By joining predicate constructions with logical connectives we obtain more complex statements. For example we formerly wrote the $|z| > 1$ without breaking the statement down into elements, but now we write it

$$> (z, \text{"}1\text{"}) \vee < (z, \text{"}-1\text{"})$$

■ QUANTIFIERS

IN MATHEMATICS a large role is played by assertions of the universality of a given attribute and of the existence of at least one object that possesses the given attribute. To record these assertions the following so-called quantifiers are introduced: universal quantifiers $\forall$ and the existential quantifier $\exists$. Let us suppose that a certain statement S contains a variable (indeterminate object x; therefore we shall write it in the form $S(x)$. Then the statement $(\forall x)S(x)$ means that $S(x)$ occurs for all x, while the statement $(\exists x)S(x)$ represents the assertion that there exists at least one object x for which the statement $S(x)$ is true.

A variable included in a statement under the sign of a quantifier is called a *bound* variable, because the statement does not depend on this variable just as the sum

$$\sum_{i=n}^{m} S_i$$

does not depend on the indexes i. The bound variable may be replaced by any other letter that does not coincide with the remaining variables and the meaning of the statement will not change as a result. A variable which is not bound is called *free*. The statement depends entirely on the free variables it contains.

Here are some examples of statements containing quantifiers.

1) $(\forall x)\,(\forall y)$ ["brother" (x,y) & "man" $(y) \supset$ "brother" (y,x)].

For every x and every y, if x is the brother of y and y is a man then y is the brother of x.

2) If $D(x,y)$ is used to represent the statement "x is a divisor of y," then one of the relationships cited above as an example of a statement will be represented in the form

$$(\forall n)\,[> (n_1 \text{ "1"}) \supset (\exists p)D\,(P_1\, n)]$$
$$(\exists x)W(x) \supset - (\forall x) - W(x)$$

The last relation is true for any statement $W(x)$ and shows that there is a connection between the universal and existential quantifiers. From the existence of object x for which $W(x)$ is true it follows that the assertion that "$W(x)$ is untrue for all x" is not true.

A quantifier is also, in essence, a logical connective. The attribution of a quantifier changes a statement into a new statement which contains one less free variable. The difference from the connectives we considered above is that one must indicate, in addition to the statement, the free variable that must be coupled. The coupling of a variable means that concrete objects will be put in its place. If the number of objects that can be substituted for the variable is finite then the quantifiers can be viewed simply as convenient abbreviations because they can be expressed by the logical connectives of conjunction and disjunction. Suppose variable x can assume n values, which we shall designate by the letters $x_1, x_2, \ldots , x_n$. Then the following equivalences will occur.

$$(\forall x)W(x) \equiv W(x_1)\,\&\,W(x_2)\,\&\ldots\&\,W(x_n),$$
$$(\exists x)W(x) \equiv W(x_1)\vee W(x_2)\vee \ldots \vee W(x_n)$$

■ THE CONNECTIVE "SUCH THAT"

THE THIRD LINE of our table describes a construction that correlates an object to a statement. In natural languages this construction is very widely used. When we say "red ball," we have in mind an object "ball" which possesses the attribute "red," that is, it is such that the statement "red" ("ball") is true. We transfer the statement about the object to the adjective which modifies the noun by which we designate an object; in other cases this can be achieved by participles, participial constructions, and constructions with the connectives "which" and "such that." If we carry this analysis further we shall find that the noun, like the adjective, indicates first of all a definite attribute or attributes of an object. Like the word "red," the word "ball" depicts a certain class of objects and may be correlated to a one-place predicate, "is a ball" (x), or simply "ball" (x). Then "red ball" is such an object that the statements "ball" (a) and "red" (a) are true; in other words, the statement "ball" (a) & "red" (a) is true.

Notice that there are three independent elements operating in the logical notation: the letter a and the objects "ball" and "red," while in writing in natural language there continue to be just two, "red" and "ball." But the letter a, which is introduced in logical notation to identify the given object and distinguish it from others (and which is called the *identifier*), does not completely disappear in natural notation. It has been transferred to the concept "ball," changing it from an *attribute* to an *object!* Unlike the word "red," the word "ball" identifies; you can say, "This is the ball we lost yesterday" or "I have in mind the same ball I was talking about in the previous sentence."

But what is an "object"?

■ THE PHYSICAL OBJECT AND THE LOGICAL OBJECT

EXPERIENCE TEACHES us that the world we live in is characterized by a certain stability and repetition (and also, of course, by constant movement and variation). Suppose we see a tree. We walk away from it and the image of the tree on the retina of our eye changes in relationship to our movements. This change follows a definite law which is very familiar to us from observation of other objects. But when we return to our former place the image becomes almost exactly the same as it was before. Then we say, "This is the tree," having in mind not only the image of the tree—the mental photograph—at the given moment in time but also the situations at nearby moments. If we are talking about classifying distinct situations by themselves, without considering their relations to other situations, then there is no difference at all between the noun and the adjective; the concept "ball," just like the concept "red," is completely defined by indicating a certain set of situations, and the discriminator (natural or artificial) of these concepts need only be able to use the following sentences correctly: "This is red," "This is not red," "This is a ball," and "This is not a ball."

It is different when we must classify time sequences of situations rather than separate situations; we shall represent them as if they were a movie film whose frames each depict the situation at a given moment. In the movie film "ball" is not simply a detail of the situation in one frame; it is a detail that recurs in many. The discriminator of the concept "ball" cannot simply say, "Yes, my friends, this is a ball!" It must identify the particular details in the frames, saying: "Here is how the ball looks in frame no. 137; here is the same ball in frame no. 138; here it is again in frame no. 139; and here is what it looked like in frame no. 120," and so on. The detail of the situation which is called "the same ball" can change quite considerably because of change in the position of the eye relative to the ball or a change in the shape of the ball itself, but the ball itself is invariably and absolutely the same. This invariability reflects the relative and temporal invariability we find in reality. It is as if we were to draw a line in time connecting the details in the different frames of film and declare that everything on this line is "the same" object. It is this line, in combination with a certain set of attributes (characteristics), that forms the concept of the physical object.

The logical concept of the object reflects a property of physical objects—they preserve their identity. The object of logic is simply an identifier. Sameness is its only attribute, as reflected in our imaginary connecting line. If there are several different classes of objects, then various types of identifiers are ordinarily used to denote the objects in different classes. For example, line segments will be represented by small letters, points by capital letters, angles by Greek letters, and so on. But more concrete attributes characteristic of objects are written in the form of distinct assertions which include the introduced designations. This makes it possible to get by without a construction involving the connective "such that." It is true that at the very beginning of his famous treatise *Éléments de mathematique* Bourbaki introduces the designation $\tau_x[A(x)]$ for a certain object which possesses attribute $A(x)$, that is, such that $A\{\tau_x[A(x)]\}$ is a true statement. After this, however, the designation disappears from his text. Thus a definite name for the construction that associates an object with a statement has not even been established and we are forced to leave a blank in our table. In the last analysis, a complete division of labor between identifiers and statements is more convenient.

For example let us take the sentence: "The reddish-brown dog of Lieutenant Pshebyssky's widow killed the stray cat." When written in the language of logic this sentence breaks down into several statements which are implicitly contained in it and expressed by means of the grammatical category of attribution. They can be joined into one statement using the conjunction sign, but we can obtain a more conventional and readable notation if we simply write out all the assertions being made—each on a new line separated by commas instead of conjunction signs. Assuming that the meaning of the attributes and relations being introduced is clear from the context, we receive the following equivalent of the above sentence:

> "dog" (a),
> "reddish-brown" (a),
> "belongs" (a, b),
> "widow" (b, c),
> "Lieutenant Pshebyssky" (c),
> "killed" (a, d),
> "cat" (d),
> "stray" (d).

■ FUNCTIONS

IN THE EXAMPLE, the predicate, "Lieutenant Pshebyssky" (c), is the only one that is plainly not elementary. In the attribute "to be Lieutenant Pshebyssky" we distinguish two aspects: to have the *rank* of lieutenant and to have the *surname* Pshebyssky. That is why this predicate is expressed by two separate words. Of course, we could have put each of these words in the form of a distinct predicate, but the fact that "lieutenant" is the *rank* of object c and "Pshebyssky" is the *surname* of it would not have been reflected in this case, and therefore we considered such a separation meaningless.

"Surname" and "rank" are examples of a function of one free variable—of a construction that juxtaposes the object which is the meaning of the function to the object which is the free variable. The function is written, as customary in mathematics, "surname" (x), "rank" (x), and so on. If there are several free variables they are separated from one another by commas, after which we are dealing with the function of several variables. This construction associates an object-value with a set of object-variables (their order is important!). An example of a function of two free variables is "the result of a game of chess" (x, y). Let us give examples of functions from mathematics. Functions of one free variable: $\sin (x)$, $|x|$; functions of two variables: arithmetic operations which may be written $+(x,y), -(x,y)$, and so on; the distance (A,B) between two points in space A and B; a function of three variables: the angle formed at point B by paths to points A and C; the designation $<(A,B,C)$, abbreviated as $<ABC$.

Not every object can be substituted into the given function as a free variable or variables. If object a is a reddish-brown dog, then obviously the construction "rank" (a) is meaningless. The construction $+(a,B)$ is also meaningless where a is a number and B is a point in space. The set of objects (or sets of groups of objects) that can be free variables of a function (or functions) is called its *domain of definition*. The domain of definition of the function "rank" (x) is formed by all those objects which are military servicemen. The objects which can be values of the given function form the set which is called the *area of values* of the function. The area of values of the function "rank" (x) includes such objects as "ensign," "lieutenant," "major," and the like, but it cannot include "3.14" or "reddish-brown dog." The function "rank" (x) ascribes a definite rank to each serviceman.

When we deal with functions, one of the relations among objects, the relation of equality, becomes particularly important. It is essential for establishing correspondence between functional constructions and the names of objects from the area of values of the function. When singling out an equality from the mass of other relations, we preserve its conventional notation $x = y$ instead of writing it in the form of a predicate $= (x, y)$. The fact that object c has the surname "Pshebyssky" and the rank "lieutenant" will look as follows:

$$(\text{"surname"} (c) = \text{"Pshebyssky"}) \,\&$$
$$(\text{"rank"} (c) = \text{"lieutenant"})$$

The equality relation can be defined formally by the following four assertions.

1. $(\forall a)(a = a)$
2. $(\forall a)(\forall b)[(a = b) \supset (b = a)]$
3. $(\forall a)(\forall b)(\forall c)[(a = b) \,\&\, (b = c) \supset (a = c)]$
4. $(\forall a)(\forall b)\{[a = b] \supset [W(a) = W(b)]\}$

The last assertion is true for any statement $W(x)$ which depends on variable x. As an exercise we suggest that the reader translate these assertions into natural language.

In one of the examples given above we introduced the predicate $D(x, y)$, which has the meaning "x is a divisor of y." The concept of divisibility is wholly determined by the operation (function) of multiplication; therefore the predicate $D(x, y)$ can be expressed by the function x. The natural (that is, whole positive) number p is a divisor of the number n when and only when there exists a natural number m such that $n = pxm$. In the language of predicate calculus

$$(\forall p)(\forall n)[D(p_1 n) \equiv (\exists m)[n = x(p_1 m)]\}$$

To each function from n free variables we may correlate an $n + 1 =$ place predicate which expresses the relation where one (for example the last) free variable is the given function from the remaining variables. For example, corresponding to function $x(x, y)$ is the predicate $M(x, y, z)$, which yields a true statement if and only if $z = x \times y$. In the general case, corresponding to the function $F(x, y, \ldots z)$ there is the predicate $F(x, y, \ldots z, u)$, which possesses the property:

$$(\forall x)(\forall y)\ldots(\forall z)(\forall u)\{F[x, y, \ldots, z, u] \equiv [f(x, y, \ldots z) = u]\}$$

The predicate F in fact expresses the same concept as the function F. Any statement which contains functional symbols can be rewritten, using predicate symbols only and introducing a certain number of additional object variables. Thus neither of the constructions that generate new objects—the construction with the connective "such that" and the function—is essential in principle, and it is possible to get along without them. Unlike the construction "such that," however, functional symbols are very convenient and they are used extensively in logic.

■ SYNTAX AND SEMANTICS

IN CONCLUDING our short sketch of logic we shall consider the question of the relation between the language of logic and natural language. In the course of our discussion the important concepts of the *syntax* and *semantics* of language will be introduced.

Let us recall the sentence about the reddish-brown dog, which we expanded into a set of statements expressed by means of predicates. The meaning of this set coincides with the meaning of the initial sentence, but the form of notation, the structure of the text, differs fundamentally. In semiotics (the science that studies sign systems) the aggregate of rules of construction of language elements is called its *syntax* and the relationship between language elements and their meanings is called *semantics*. Thus, the first thing that strikes one's eye in comparing logical and natural language is that the language of logic has a different syntax—one that is simple and uniform. It is based on the style of notation which has taken shape in mathematics; the construction of more complex language elements from simpler ones is represented by analogy with mathematical notation of operations and functions. The syntax of the language of logic is *completely formalized*, that is, there is a set of precisely formulated rules with which one can construct any language element. Moreover, no matter what correctly constructed language element (object or statement) we may take it will always be possible to re-create the element's construction. This process is called *syntactical analysis* of the element. It is easy to see that syntactical analysis is extremely simple and unambiguous in the language of logic.

The syntax (in the semiotic sense) of natural language is its grammar—that is, the rules by which sentences are constructed from words (syntax in the narrow, linguistic sense of the word) and the rules for constructing words from letters (morphology). Unlike the language of logic, the syntax of natural language is far from completely formalized. It includes an enoromous number of rules with an enormous number of exceptions. This difference is entirely understandable. The language of logic was created artificially, while natural language is a result of long development which no one controlled consciously, in which no preconceived plan was used. The grammar of natural language has not been constructed or designed; it is an investigation of an already complete system, an attempt to discover and formulate as clearly as possible those rules which speakers of the language use unconsciously.

Syntactical analysis of the sentences of natural language often requires reference to semantics, for without considering the meaning of a sentence it will be ambiguous. For example, let us take the sentence: "Here are the lists of students that passed the physics exam." In this sentence the attribute "that passed the physics exam" refers to students. If we use parentheses to make the syntactical structure of the sentence more precise, as is done in writing algebraic and logical expressions, they would be placed as follows: "Here are the lists of (students that . . . passed)." Now let us take the following sentence: "Here are the lists of students that were lying on the dean's shelf." Formally the structure of this sentence is exactly the same as in the preceding one. But in fact a different placement of parentheses is assumed here, specifically: "Here are the (lists of students) that . . . were lying." When we arrange parentheses in this way mentally we are relying exclusively on the meaning of the sentence, for we assume that students could not be lying on the dean's shelf.

In general, constructions with the word "that" [Russian *kotoryi*] are very treacherous. In his book *Slovo o slovakh* [A Word about

Words], L. Uspensky tells how he once saw the following announcement:

CITIZENS—TURN IN SCRAP MATERIAL TO THE YARDKEEPER THAT IS LYING
AROUND!"

It is not surprising that this construction did not find a place in mathematical logic!

■ LOGICAL ANALYSIS OF LANGUAGE

THUS, to make a logical analysis—to construct the logical equivalent of a sentence in natural language, we must first of all make a syntactical analysis of the sentence; the results of the analysis will be reflected directly in the syntactical structure of the logical expression. But semantics is by no means left out either. When we say that "reddish-brown" is a one-place predicate, "killed" is a two-place predicate, "distance" is a function, "and" is a logical connective, "all" is a quantifier, and so on, we are performing a semantic analysis of the concepts expressed in natural language. We classify concepts in accordance with a scheme fixed in the language of logic and we establish relations among the concepts. Logical analysis is essentially semantic analysis. Syntactical analysis is necessary to the extent that it is needed for semantic analysis.

Logical analysis may be more or less deep. In our example it is very superficial. Let us see if we could continue it, and if so how.

The concepts of "dog," "reddish-brown," and "cat" are one-place predicates, obviously elementary and not subject to further analysis. These are simple Aristotelian concepts which rely directly on sensory experience; every normal person is able to recognize them and the only way to explain what a "cat" is is to point one out.

The concept "stray" is also a one-place predicate, although a more complex one. If we were reasoning in a formal grammatical manner we could conclude that "stray" means "that which strays." But this would be an incorrect conclusion because the verb to stray does not designate a length of time. A perfectly well-bred house cat may go out for an hour and stray across a roof, but this does not give anyone the right to call it a "stray." It would be more correct to define a stray cat as a cat that has no master or, using a relation which is already included in our logical expression, as a cat that does not belong to anyone. Here is a formal notation of this definition:

$$\text{"stray" } (x) \equiv -(\exists y) \, [\text{"belongs" } (x, \, y)]$$

(It is assumed here that x is an arbitrary object.)

Let us look at the relation "belongs." In a certain sense we sneaked it in, because the word "belong" was not in the initial sentence. But it was understood, and semantic analysis revealed it! In the Russian sentence the relation of belonging was conveyed by the genitive case. Here we see a clear example of the ambiguity and inadequacy of syntactical analysis. We used the genitive case in the constructions "the widow's dog" ["the dog of the widow"] and "the dog's mistress," but in no way can it be said that the mistress *belongs* to her dog. The construction "the widow's nose" can, of course, be interpreted as the "nose that belongs to the widow." But here we are already encountering the semantic ambiguity of the word "belong," for it is obvious that the nose belongs to the widow in a different way than the dog belongs to her.

It requires a good deal of work to break the concept "belong" into its elementary constituent parts; this would require a description of the customs and laws related to the right of ownership. Only in this case can the meaning of "belong" be explained. The predicate "widow" and the functions "rank" and "surname" (which we introduced during our analysis of the concept "Lieutenant Pshebyssky") are also bound up with the social sphere and require further analysis. Finally, the concept "killed," although it is not linked with the social sphere and is plainly simpler (closer to sensory experience) than the preceding concepts, would also have to be subjected to logical analysis. In this analysis it would be possible to identify, first of all, the element of completion in the action, which is expressed by the Russian verb form, secondly the final result (the death of the victim), and thirdly, the typical characteristic of the action expressed by the Russian verb [*zagryzt'*]—use of the teeth.

Logical analysis of language is an extremely interesting line of investigation, but we cannot dwell on it here. Those who are interested are referred to *Elements of Symbolic Logic* (New York: Free Press, 1966) by H. Reichenbach, one of the founders of this field.

Let us summarize the results of our comparison between natural language and the language of logic. The language of logic has a simple and completely formalized syntax. By syntactical and semantic analysis a text in natural language can be translated to the language of logic—that is, it can be correlated with a text in the language of logic that has the same meaning. Semantic analysis of the natural text during translation may be more or less deep, which is to say that the predicates and functions included in the logical text may be closer or further from immediate sensory and emotional experience. There are predicates and functions which cannot be broken down into more elementary constituents and which therefore cannot be defined in any way except by reference to experience. We shall call such predicates and functions *primary*.

CHAPTER SEVEN
Language and Thinking

■ WHAT DO WE KNOW ABOUT THINKING?

THE FIRST THING we must do to approach the problem of language and thinking correctly is to clearly separate what we know about thinking from what we do not know. We know that thinking is a process that takes place in the nerve nets of the brain. Because the term "representation" to us means a state of some subsystem of the brain it may be said that thinking is the process of change in the aggregate of self-representations. But at any given moment in time only a certain (obviously small) part of these representations is accessible to, as we say, our *consciousness*. These representations can be consolidated into one (for several subsystems taken together constitute a new subsystem), which is the state of consciousness at the given moment. We do not know what consciousness is from a cybernetic point of view; we have only fragmentary information (specifically, that consciousness is closely related to the activity of what is called the *reticular formation* of the brain).

Thus, thinking has an external, manifest aspect: a stream of conscious representations. This stream can be fixed and studied, and from it we try to draw conclusions indirectly about those processes in the brain which are illuminated by consciousness. We are fairly sure about some things regarding the stream of consciousness. We know that it is regulated to a significant degree by associations of representations which form under the influence of experience and reflect the characteristics of our environment. Specifically, we receive our ability to foresee future situations to one degree or another thanks to the association of representations. We also know that humans, unlike animals, have the ability to control the process of association; this is manifested as imagination, encoding, and conscious memorization. But we do not know the concrete cybernetic mechanism of this ability or, as a matter of fact, the mechanism of the association of representations. These mechanisms are not given to us subjectively either; in the stream of consciousness we merely observe their appearance, the result of their action. Finally, we are subjectively given a sensation of freedom of choice in our actions: free will. Free will also manifests itself in thinking. We are able to turn our thoughts to any subject we wish. We do not know the cybernetic interpretation of free will either, and this situation is perhaps worst of all.

■ LINGUISTIC ACTIVITY

REPRESENTATIONS of linguistic objects, words and sentences, occupy a distinct place among all representations in the process of thinking. These representations are (with the exception of deaf mutes, of course) a combination of aural and motor representations and (for people who have dealt with written language from childhood) the visual component may also be joined to them. When we picture a certain word in our mind we mentally pronounce it, listen, and possibly see it written. For brevity we will call these *linguistic representations*. The stream of linguistic representations is precisely what is ordinarily called thinking. The presence of this stream is a specifically human characteristic; it is not found in animals. So-called "abstract" thinking is actually thinking in words, the stream of linguistic representations. Without such thinking, the achievements of thought in which the human race takes such pride would have been impossible.

The significance of linguistic representations is that they are uniquely related to words and sentences as the material elements of the material system "language." This system is the aggregate of all words and sentences pronounced orally, transmitted by telephone and radio, written on paper, encoded on punched cards for computers, and so on—in short, the aggregate of what we have called the higher nervous system of the material body of culture. Functionally, a stream of linguistic representations in no way differs from a sequence of their material correlatives: words. The external, observed aspect of thinking may be described as activity consisting of the creation of certain material linguistic objects, for example pronouncing sentences out loud (unfortunately these objects are very short-lived) or writing them on paper. We shall call this activity *linguistic*.

There are compelling reasons to consider linguistic activity the basic, primary aspect of thinking and the stream of linguistic representations merely a transitional element—a form of connection between the material linguistic objects and the aggregate of all (not just linguistic) representations. In fact, it is precisely the linguistic objects which store and transmit information and operate as the elements of linguistic models of reality. The child is taught linguistic activity in the same way as it is taught to walk, shoot a bow, or hammer nails. As a result the child becomes, so to speak, plugged into the language; he uses the models already available and enriches it with new ones. Furthermore, he may also use language in a noncommunicative manner (for his own purposes) as did the young man Uu of the Nyamnyam tribe when he counted the enemy with his fingers. During noncommunicative use of language there may be a stream of linguistic representations without apparent linguistic activity ("I think!"); but after all, these representations emerged and acquired their meanings as a result of activity involving substantial, material linguistic objects! And often during the process of reflection we whisper certain words and whole phrases, returning them to their material form. The primacy of substantive linguistic activity is especially clear when we are dealing with scientific models of reality. After long, hard study with real, written symbols a person may be able to multiply a few small numbers or reduce similar elements of an algebraic expression in his head. But give him a problem that is a little harder and he will demand a pencil and paper!

Linguistics and logic investigate linguistic activity. Linguistics is interested primarily in the syntax of language (in the broad, semiotic sense) while logic is chiefly interested in semantics. When syntax and semantics are interwoven it is not possible to separate linguistics from logic. It is true that traditional logic declares itself to be the science of the laws of thinking, not the science of language, but this pretentious statement should not be taken too seriously. Of all the fields of knowledge which study thinking, logic has the most external, superficial approach. It does not investigate the real mechanisms of the work of the brain, as neurophysiology does; it does not construct models of mental activity, as cybernetics does; and it does not attempt to record and classify subjectively perceived emotional states, as psychology does. It recognizes only precise, socially significant thoughts (not the ravings of a madman!) as its object of study. But such thoughts are in fact nothing else but linguistic representations with socially significant semantics. Logical (semantic) analysis of language leads to primary, undefinable concepts and stops there; it does not take us beyond language. Logic also contains its theory of proof. If language is used in a form of notation which keeps within the rules of predicate calculus, not in the form of natural language, it is possible to establish the formal characteristics of the correctness of deductions and formal rules which, if used, will always yield correct conclusions from correct premises. These rules (the laws of logic), which are also expressed in the form of a linguistic object, form a metasystem in relation to the statements obtained as a result of application of the rules (see figure 7.1). Sentences are the object and result of work for the theory of proof. Thus, all of logic lies wholly in the sphere of linguistic activity. Its lower stage is semantic analysis and its higher stage is the theory of proof. We will talk about proof theory later; for now we are interested in the lower stage (it may even be called the foundation): the relationship between language and the working of the brain.

We shall consider that by logical analysis we can translate any sentence in natural language into the language of logic. Of course, this somewhat exaggerates the advances made to date, but it is fairly clear that in principle there is nothing impossible about it. Logical analysis reveals the internal structure of language, the fundamental nodes of which it consists. Therefore we shall review the basic con-

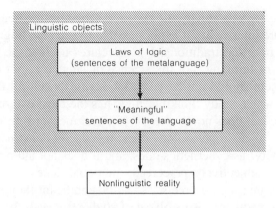

Figure 7.1. Logic as a metasystem.

cepts of the language of logic, clarify exactly why they are as they are, and discover how they are related to brain activity. Whereas in the last chapter we were primarily concerned with the syntax of language, here we shall pose the question of the semantics of language.

■ THE BRAIN AS A "BLACK BOX"

FIRST LET US try to find direct correlatives of language elements in brain activity. The first concept we introduced in our description of the language of logic was the statement. With what can it be correlated? The answer suggests itself: the association of representations. Indeed, like the brain, language is a system used to create models of reality. In the case of the brain the basic independent unit that can operate as a model is the association of representations, while in the case of language it is the statement.

Now there is a temptation to correlate the representation to the object. At first glance this creates a complete and harmonious interpretation: the object corresponds to the representation; the relation among objects, which is the statement, corresponds to the relation among representations, which is the association. We may take the example of the association "In the forest there are wolves," which we gave in chapter 4, and interpret it as follows: "forest" and "wolves" are objects and, at the same time, representations, while "In the forest there are wolves" is a statement and, at the same time, an association.

But a careful analysis shows that this interpretation involves a serious mistaken assumption; we have artificially transferred linguistic structure to the sphere of representations. In reality this sphere has no such structure. Begin from the fact that an association of representations is also a representation. A representation may be correlated with the sentence "In the forest there are wolves" just as it may be correlated with the nouns "forest" and "wolves." We should recall that an association between representations S_1 and S_2 is a new synthetic representation U (see figure 3.8). It is true that the association of representations is a model of reality, but if we understand the term "model" in the broad sense as a certain correlative of reality, any representation is a model. If, however, we understand model in the narrow sense as a correlative of reality which permits us to predict future states, then not any association can be a model, but only one that reflects the temporal aspect of reality. The process of associating is important, because it leads to the creation of a new model where none existed before. This process permits completely strict logical definition and can be revealed by experiment, similar to the way we easily define and uncover the process of the formation of a system from subsystems. But it is impossible to define the difference between an association of representations and a representation just as it is impossible to establish criteria that would distinguish a system from subsystems.

So the statement elicits a representation and the object elicits a representation and our harmonious system crumbles. The representation proves too broad and too indefinite a concept to be made the basis of a study of the semantics of language. All we know about the representation is that it is a generalized state of the brain, but we know virtually nothing about the structure of the brain.

In chapter 4 we defined language as the aggregate of objects L_i each of which is the name of a certain object R_i, which is called its meaning. Concerning objects R we said only that they are some kind of real phenomena. The time has now come to work toward a more precise answer as to what kind of phenomena these are; in other words, the question is "what are the semantics of natural language?"

In the simplest examples usually given to illustrate the relationship L_i-R_i, and which we cited above (the word lion—the animal lion, and so on), the object R_i is a representation of a definite object. In general, language emerges as the result of an association between linguistic and other representations, and therefore it is natural to attempt to define the semantics of language by means of those representations which emerge in the process of linguistic activity. It can be said that the meaning of a linguistic object is that representation which it evokes—the change in the state of the brain which occurs when a representation about a linguistic object appears in the consciousness. This definition is entirely correct, but unfortunately it is unproductive because the states of the brain as objective reality are not directly accessible to us, and we make our judgments about them on the basis of their manifestation in human actions only.

Therefore let us take another route. We shall view the brain as a black box; we shall investigate the observed manifestations of its activity without any attempt to understand its internal organization. We are interested in the semantics of language, the connection (associations) between linguistic representations and all others (see figure 7.2). Because the representations are inside the "black box," however, we shall rely only on the input data corresponding to them— which is to say the linguistic objects and all the other activity that, for the sake of brevity, we shall call nonlinguistic. This is the input of the black box. Its output is obviously the person's observed actions.

Because the system of actions is very complex, we shall not make progress in our attempts to study semantics if we do not choose some simple type of action as a standard. Of course there must be at least two variants of the action so that it will carry some information. Suppose there are exactly two. We shall call them the first and second *standard actions*. We shall formulate the elementary act in studying semantics as follows. Linguistic objects will be presented to a person who is perceiving a definite nonlinguistic reality and we shall assume that he responds to them by performing one of the two standard actions.

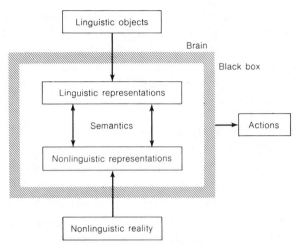

Figure 7.2. The brain as a black box.

■ AFFIRMATION AND NEGATION

WE CONCEIVED this scheme in a purely theoretical manner as the simplest method of defining the semantics of language under conditions where the brain is pictured as a black box. It turns out that this scheme actually exists in linguistic activity, emerging spontaneously in the early stages of the development of language! In all known languages we find expressions for two standard actions—*affirmation* and *negation*. These actions are of great antiquity, as evidenced by the fact that among a large majority of peoples (possibly all) they are expressed in gestures as well as words. If we open the top of the black box just a crack, to the degree shown in figure 7.2, we can define the affirmation as an action performed when the linguistic object and reality are in the relation *name-meaning* (that is, the necessary association exists between the linguistic and nonlinguistic representations), and we can define negation as the action performed when there is no such relation. But a person learning to use affirmative and negative words and gestures correctly knows nothing, of course, about representations, associations, and the like. At first he is simply taught to say "cat," "dog," and so forth while pointing at the corresponding objects, and then he is taught to perform the affirmative action when someone says "this is a cat" while pointing at one and to perform the negative action when someone makes the same statement while pointing at a dog. In both instances we learn correct linguistic activity while relying on the brain's ability to recognize and associate; but we have no knowledge of the brain's mechanisms; to us it is a black box.

The last remark explains why it is hardly surprising that the scheme of standard actions has become an established part of linguistic practice. A person's brain is a black box both for himself and for other members of society. This is the origin of the need for a socially meaningful way of determining more precise semantics; this need appears as soon as language reaches a minimum level of complexity.

The standard actions of affirmation and negation are not related to reality itself, as primary linguistic objects are; rather they refer to the relationship between primary linguistic objects and reality. They are elements of a metasystem in relation to the system of primary linguistic objects. The introduction of the actions of affirmation and negation into the practice of society was the beginning of that metasystem transition within linguistic activity whose subsequent stages are the appearance of the language of logic and the theory of deduction. Although affirmation and negation appeared very early in the development of human culture, they did not appear sufficiently early for a prototype of them to be found in animal actions. We know that such prototypes exist for primary linguistic objects in the form of animal signals. Among these signals there are ones which could be described as affirmative and negative, but they have nothing in common with the semantic actions of affirmation and negation which are oriented to the signals themselves and lay the foundations of the metasystem. In this we see one more manifestation of the law of branching (expansion) of the penultimate level. The enormous growth in the number of primary linguistic objects (signals) which is found in human society began simultaneously with the beginning of the metalevel.

■ THE PHENOMENOLOGICAL DEFINITION
OF SEMANTICS

NOW IT WILL not be difficult for us to interpret the basic concepts of logic from the point of view of the phenomenological ("black box") approach. The statement is obviously the linguistic object to which the actions of affirmation and negation refer. The semantics of a language appear to an external observer as the function of two free variables (the statement and the true state of affairs); the function assumes one of two truth values: "true" ("yes," "truth") and "untrue" ("no," "falsehood"). The value of this function is worked out by the black box, the human brain, which knows the given language. How this happens the external observer does not know.

The statement is the basic unit of language. In considering language as a system we must discover how the statement, a system of statements of subsystems, can be constructed. Thus we come to the introduction of *logical connectives,* which were discussed in the preceding chapter.

Reality is perceived by the human being through the medium of the sense organs; it appears to the human being as an aggregate of receptor states, a *situation.* If a person were unable to control his sense organs and concentrate his attention on certain parts of the situation, that is, if the situation always appeared to a person as something whole and completely given from outside, then all logic would probably be limited to propositional (statement) calculus. But a person can control his sense organs and can, for example, fix his vision on a particular object. Therefore the situation is not simply reality, it is reality with an *attention characteristic*—that is, with an isolated area (approximately defined) which we are speaking about and on which we concentrate our attention. The concept of attention also has a psychological aspect, but we shall try to bypass it. We can determine from observing a person what he is looking at (or feeling, smelling, and so on), because the attention characteristic can be determined objectively. Reality with the attention characteristic can therefore be viewed as a free variable of the function in the "black box" approach. People resort to gestures or verbal clarifications to define the position of the area of attention more precisely. In either case the result will be the same. If you say, "I am looking at the thick book the girl in the pink dress is holding in her hands," the person you are talking to will look around until he locates the girl and the book.

The temporal aspect of the input data of semantics must also be taken into account. If the reaction of the brain were determined only by the situation at one specific moment, unrelated to situations close in time, once again logic would probably be limited to propositional calculus. In fact, however, the brain stores its memory of many past situations; the brain's reaction (and specifically, the standard action) is therefore always a function of the *moving picture* of situations. We often fail to recognize this because there are in the environment around us objects which show a relative invariability, and when we concentrate our attention on the invariant object it seems to us that we are not dealing with a moving picture but rather with a single frame. In actuality, the analysis of the concept of object which was given above shows that the time aspect plays the decisive part in it. Now, when we have introduced the concept of the attention characteristic we can define the object as a moving picture of situations with the attention characteristic represented by one continuous line.

The extent to which we are inclined to ignore the dynamic aspect of perception can be seen from the situation we ordinarily describe as the existence of at least two distinct objects. It seems to us that we are perceiving each object separately and still we distinguish among all the objects and concentrate our attention on them simultaneously. But the simplest psychological self-analysis will persuade us that in fact in such a case our attention darts rapidly from one object to another. In the moving picture of situations, the line of the attention characteristic will be broken; it will, indeed, easily become possible to make several (according to the number of objects) continuous lines (see figure 7.3).

We have now come to defining the concept of the object in logic. We have established that the "nonlinguistic activity," shown in 7.2, which is fed to the input of the black box is often broken, divided up in space and time. It can be imagined as a moving picture

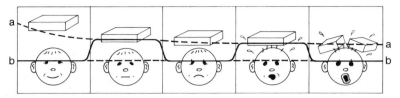

Figure 7.3. Broken line of attention out of which two continuous lines can be formed.

on which the line of movement of the attention characteristic is drawn in. Moreover, it turns out that this line can be broken to become several continuous lines. These continuous lines are the objects.

Thus the object of logic is wholly liberated from its material meaning; this is transferred to statements about the given object. The object is an identifier. Its only attribute is to be identical to itself and it signifies a continuous line of attention. This proposition has already been explained in sufficient detail in the preceding chapter.

When in place of undivided reality we feed to the input of the black box a reality divided into objects, the statement becomes dependent on the method of division—that is, on the objects we are singling out; the statement is converted into the *predicate*.

■ THE LOGICAL CONCEPT

WE HAVE ALMOST completed our analysis of the fundamentals of logic from the black-box point of view. We have still to define the general concept of "logical concept," but that is simple: the concept is the predicate or logical connective. The grounds for this are that predicates and connectives are those basic functional nodes we discover in linguistic activity. The concept of function in the sense that we have defined it above may not be elevated to the rank of the basic logical concept because, as we have seen, it can be expressed through predicates and connectives. But in the broader sense both logical connectives and predicates are functions—that is, correlations by a certain method of values (truth values in the given case) to free variables. Thus it can be said that the logical concept is a function whose free variables are linguistic objects and situations and whose values are linguistic objects. The result of a logical analysis of language is a breakdown of linguistic activity into homotypic functional elements: connectives and predicates.

Every logical concept is defined in the first place by its material carrier, the linguistic object (in most cases a word or phrase), and in the second place by the method of using this object in linguistic activity in society. The second point offers an opportunity to refine the first. The words "koshka," "koshka," "KOSHKA," and koshka [Russian for "cat"] are different linguistic objects (the first two differ by their placement while the third and fourth also differ by their type face) but we consider them to be carriers of a single concept because they appear indistinguishable in linguistic activity. The same thing can be said—with certain restrictions—about the German die Katze, for it is used analogously (but only analogously!) to the Russian koshka.

The concepts of a language form a hierarchical system. In certain specialized languages (sublanguages) used by the exact sciences this hierarchy is determined in a completely clear and strict manner. The concepts located higher in the hierarchy acquire their meaning by logical definitions through concepts of a lower order—that is, it is pointed out how, being able to determine the truth values of the predicates at a lower level, one can determine the value of the predicate of a higher level. In natural languages there is no strict hierarchy, but there is an approximate one. We can therefore assess the "degree of remoteness" of a concept from the direct data of experience by logical analysis and breaking complex concepts down into simpler components; the degree of remoteness of a concept from direct experience can be equated with its elevation in the hierarchy. This estimate of position in the hierarchy is approximate, because the breakdown in the components is not unambiguous, the actual method of subdividing has not been fully formalized, and no one has yet done such work for all language. Perhaps the most firmly established fact is that the predicates which cannot be subdivided at all are primary (belonging to the lowest level of the hierarchy).

Between the concepts of a language there are numerous interconnections given by the set of all true statements in which the concepts under consideration are included. Language is a system and its concepts have meaning only as elements of the system. The meaning of a word is determined by the way this word is used in linguistic activity. Each word, so to speak, bears the imprint of all the sentences in which it has ever been included; it is an element of the system. When traditional logic speaks of concepts, the two functions of the concept are pointed out: to serve as an element of reasoning—that is, a method of shaping thought—and at the same time to concentrate already existing thoughts and knowledge of an object in oneself. This duality is a result of the *system nature* of the concept. The linguistic object (word) which expresses a concept is used as an element for constructing a model of reality and is associated functionally—that is, in linguistic activity (and therefore also in our imagination)—with all models in which it participates. Therefore, although a trained dog does distinguish between a square and a circle it cannot be said that it has mastered the concept of "square"; this word includes many things about which the dog does not have the slightest idea. Therefore also the most exact translation from one language to another is by no means always a literal translation; the difference between the systems must be taken into account. Strictly speaking, an absolutely exact translation is generally impossible (with the possible exception of statements which contain only primary concepts accessible to a dog).

■ THE STRUCTURAL APPROACH

WE HAVE DEFINED the logical concept as an element of the functioning of the linguistic system. We shall now attempt to give a more general definition of the cybernetic concept of "concept," relying on the structural rather than the functional approach.

Let us again consider the concept "inside" in application to the picture discriminator. How would we begin to build a system that contains the concept "inside"? It is apparent that at first we would have to construct classifiers for the concepts of "spot" and "contour." Let us recall that the classifier is a cybernetic system that recognizes the affiliation of an input state ("situation") with a definite set (Aristotelian concept), and converts it to an output state that reflects the most important characteristics of the situation. The spot classifier, for example, recognizes the existence of a spot and fixes the coordinates of the points which bound it. In figure 7.4 we have designated the classifiers of spots and contours by the letters π_1, π_2, . . . and K_1, K_2, . . . These classifiers form the first level of a hierarchy, for their input is the states of the receptors. They translate the situation from the language of illuminated points to the language of spots and contours.

Having constructed the first level, we begin work on the second. We construct classifier B (see figure 7.4) to whose input is fed the output of one spot classifier, π_i we shall assume, and one contour classifier, K_j. Classifier B must have just two output states: one ("yes") occurs when the spot fixed by classifier π_i lies inside the contour fixed by classifier K_j, while the second ("no") occurs in the opposite case. We would like classifier B to be applicable to any pair (π_i, K_j). But it would be insanity to make as many copies of B as there are pairs (π_i, K_j)! therefore we need some kind of switching device by means of which information from different points of the system could be fed to the one and only device B. Because it is meaningless to feed information directly from the receptors or from any other inappropriate points to a classifier, the switch should be designed so that it is able to feed information from any of the pairs (π_i, K_j) and nothing else.

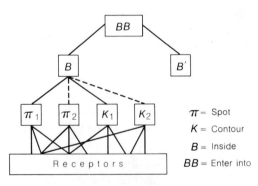

Figure 7.4.

154 LANGUAGE AND THINKING

Classifier B is located on the second level of the overall system. It may possibly be used as an input for the third level. For example, let us suppose that the system is required to recognize the concept "enter into . . ." This is a dynamic concept related to time. As the input here we must consider not one situation but rather a series—that which above was called a moving picture of situations. With such a moving picture we say that the spot has "entered into" a contour if at first it was outside the contour and then assumed a position inside it. It is apparent that the discriminator of the concept "enter into" (in figure 7.4 it is designated BB) will require at its input the output from discriminator B or from several discriminators B' related to different frames of the moving picture (in the first case it should have a device for storing the sequence of "yes" or "no" answers).

A hierarchy of classifiers has been obtained. For us this is not new; in chapter 2 we considered hierarchies of classifiers. But in that chapter we limited ourselves to Aristotelian concepts, and the hierarchy of classifiers acted solely as a means of recognizing concepts and was not included in the definition of the concept of the "concept." We defined the concept of the "concept" (Aristotelian) independently of the organization of the hierarchy of classifiers as a certain set of situations—in other words a function that assumes a truth value of "true" in the given set of situations.

But now, searching for a cybernetic interpretation of such concepts as "inside," we see that we cannot define the more general concept of "concept" by relying on the level of receptors alone; instead it can only be defined as an element of a *system of concepts*. Corresponding to the concept of "inside" in figure 7.4 is the classifier B, not only as a device which converts the given input into the given output but also as a subsystem of the total recognition system— that is, as an element connected in a certain way with other elements of the system (in the particular case, receiving input information from one type π classifier and one type K classifier).

We have constructed a cybernetic model of the concept "inside." But how is this model related to reality? What relationship does it have to the true concept of "inside," which manifests itself in language and appears to us as one of the elements of our thinking?

Can it be asserted that the brain has a classifier that corresponds exactly to the concept of "inside"?

Although the general appearance of the diagram in figure 7.4 with its receptors and classifiers reflects neurophysiological findings, the concrete functions of the classifiers and their interrelationships reflect logical data. Therefore our diagram is not a model of the organization of the brain, but rather a model of the functioning of the linguistic system—or more precisely a structural diagram of a device that could perform the functions discovered in linguistic activity. In this device the classifiers perform the functions described by logical concepts and the switching devices (which are not shown in the diagram but mentioned in the text) fix the domain of definition of the concepts.

The diagram shown in figure 7.4 may be embodied in a real cybernetic device whose sources of information will be the illuminated points of a screen. But even if such a device works very well it will not, strictly speaking, yet give us the right to consider it a model of the organization of the brain. Possibly the division of the nerve nets into classifiers as suggested by figure 7.4 or analogous diagrams taken from the functioning of language does not reflect the true organization of the brain at all!

■ TWO SYSTEMS

WE HAVE BEFORE US two cybernetic systems. The first system is the human brain. Its functioning is individual human thinking. Its task is to coordinate the actions of separate parts of the organism in order to preserve its existence. This task is accomplished, specifically, by creating models of reality whose material body is the nerve nets and which we therefore call neuronal models. We know that the brain is organized on the hierarchical principle. We call the structural elements of this hierarchy classifiers. The functions of the classifiers, considering their systems aspect—which is to say their interrelationships—are the individual concepts (in the cybernetic sense of the word, which simply means according to the cybernetic definition of the concept of "concept"), which may be identified in the functioning of the brain as a whole. We will call them *neuronal concepts*.

The second system is language. Its functioning is linguistic activity in society. Its task is to coordinate the actions of individual members of society in order to preserve its existence. This task is accomplished, specifically, by creating models of reality whose material body is linguistic objects and which we therefore call *linguistic models*. Like the brain, language is organized hierarchically. The functional elements of this hierarchical system are the *logical (linguistic) concepts*.

These systems are by no means independent. The linguistic system is set in motion by the human brain. Without the brain, language is dead. On the other hand, the brain is strongly influenced by language.

Now the problem may be formulated as follows: what is the relationship between neuronal and logical concepts? Let us survey the sources of information about these systems of concepts. Logical concepts are on full display before us; phenomenologically speaking, we know virtually everything that can be known about them. We know very little about neuronal concepts. Neurophysiological research offers some information about the lowest levels of the hierarchy only; about the higher levels we have absolutely no information which is independent of language. But we do know that language is an offspring and, in a certain sense, a continuation of the brain. Therefore a close relationship must exist between the highest stages of neuronal concepts and the lowest stages of logical concepts. After all, logical concepts came from somewhere! The logical concept of an object unquestionably has a very definite neuronal correlative; that is, long before the appearance of language and independent of it the world presented itself to people (and animals) as an aggregate of objects. From the ease with which people and animals recognize some relations among objects (in particular transformations in time) we may conclude that there is also a special neuron apparatus for relations among a small number of objects. It can scarcely be accidental that the languages of all people have words that signify the objects surrounding human beings and words for the simplest relations among them—such as the relation of "inside," which we used as an example above. Thus figure 7.4 can be considered a model of brain organization with a certain probability after all!

When speaking of neuronal models and concepts we have in mind not only the inborn foundation of these concepts but also those concrete concepts which form on this foundation through the action of the stream of sensations. In higher animals and human beings the formation of new concepts as a result of association of representations plays an enormous part, as we know. It begins from the moment the individual appears on earth and develops especially intensively at a young age, when the conceptual "flesh" fills out the congenital conceptual "skeleton." This introduces a new element into the problem of the mutual relations of neuronal and logical concepts. Those initial neuronal concepts which form in a baby before it begins to understand speech and talk can be considered independent of language, and then logical concepts can be considered reflections of them. But the more complex concepts form in a baby under the direct and very powerful influence of language. The associations of representations which make up the basis of these concepts are dictated by society's linguistic activity; to a significant degree they are thrust upon the child by adults during the process of teaching the language. Therefore, when we analyze the interrelations of linguistic activity and thinking and attempt to evaluate the degree to which the language is a continuation of the brain we cannot view neuronal nets as a given against which the logical concepts of the particular language should be compared. Considering the inverse influence of linguistic activity on thinking, the question can only be put as follows: what would the neuronal and logical concepts be like if the development of language were to follow this or that particular path?

■ CONCEPT "PILINGS"

THE INFORMATION capacity of the brain is incomparably greater than that of language (in the process of speech). Language does not reflect the full wealth of sensations and cognitive representations. We know, for example, that the ancient Greek language had just one word for both dark blue and green; as a result they had just one concept in place of our two. Does this mean that they perceived color differently? Of course not. The human eye distinguishes hundreds of nuances of color but only a few words exist to denote them.

The primary logical concepts may be compared with buttresses or, better, with pilings driven into the ground of the neuronal concepts. They penetrate to a certain depth and occupy just a small part of the area. Floor by floor the entire building, the hierarchy of concepts of the language, is erected on these piles. We take pride in the building because it contains concepts which were not even conceived of at ground level, among the neuronal concepts. But have the pilings been driven well? Could they have been driven at other points and is it too late now to drive additional ones? How does this affect the building? In other words, is the selection of primary predicates fundamental for the development of language, culture, and thinking? We rarely ask ourselves this question because we do not see the ground itself; it is covered by the edifice of language. But if we go down under the floor we can touch the original soil and feel around in the darkness with our hands. By doing this we may learn once again how much of the ground is not touched by the pilings (especially in the sphere of spiritual experience) and we shall recall the words of the poet Tyutchev: "The thought expressed is a lie."

From this metaphor one more question arises: how good is the architecture of the building? Is it the only possible architecture, and if not, how much does its selection influence the functioning of the edifice, the possibility of expansion, remodeling, and so on? In other words, is the grammar of language (at least in its most important, fundamental features) something external and unimportant for thinking, or does it fundamentally affect thinking and direct its development?

We have formulated both of these questions, concerning the effect of selection of primary predicates and of grammar, in a form requiring a yes or no answer only for purposes of clear presentation. The point is not, of course, to answer them simply by yes or no. The answer will always, in the last analysis, contain a conditional element, and the fact that there is some influence is undoubted. Our job is to investigate real findings regarding language's influence on thinking.

■ THE SAPIR-WHORF CONCEPTION

THE WORK of two American linguists, E. Sapir and B. Whorf, is very interesting from this point of view. The following quote, which Whorf used as the epigraph to his article "The Relation of Habitual Thought and Behavior to Language," gives an idea of Sapir's views:

> Human beings do not live in the objective world alone, nor alone in the world of social activity as ordinarily understood, but are very much at the mercy of the particular language which has become the medium of expression for their society. It is quite an illusion to imagine that one adjusts to reality essentially without the use of language and that language is merely an incidental means of solving specific problems of communication or reflection. The fact of the matter is that the "real world" is to a large extent unconsciously built up on the language habits of the group. . . . We see and hear and otherwise experience very largely as we do because the language habits of our community predispose certain choices of interpretation.[1]

B. Whorf takes this conception as his basis and gives it concrete form in his studies of certain Indian languages and cultures and his comparisons of them with European languages and culture. We will present some of Whorf's observations and thoughts on such logical categories as space and time, form and content.[2]

Whorf notes that to correctly evaluate such categories one must first reject those views regarding the interaction of language and thought which are ordinarily considered an integral part of "common sense" and are called, by Whorf, "natural logic." He writes:

---

[1] Quoted from *Novoe v lingvistike* (New Developments in Linguistics), No 1, Moscow, 1960. [Original Whorf article in *Language, Culture, and Personality,* Menasha, Wisconsin, 1941, pp 75–93.]

[2] I have taken the quotes by Whorf from the above-mentioned Soviet publication, *Novoe v lingvistike.*

Natural logic says that talking is merely an incidental process concerned strictly with communication, not with formulation of ideas. Talking, or the use of language, is supposed only to "express" what is essentially already formulated non-linguistically. Formulation is an independent process, called thought or thinking, and is supposed to be largely indifferent to the nature of particular languages. Languages have grammars, which are assumed to be merely norms of conventional and social correctness, but the use of languages is supposed to be guided not so much by them as by correct, rational, or intelligent *thinking*.

Thought, in this view, does not depend on grammar but on laws of logic or reason which are supposed to be the same for all observers of the universe—to represent a rationale in the universe that can be "found" independently by all intelligent observers, whether they speak Chinese or Choctaw. In our own culture, the formulations of mathematics and of formal logic have acquired the reputation of dealing with this order of things, i.e., with the realm and laws of pure thought. Natural logic holds that different languages are essentially parallel methods for expressing this one-and-the-same rationale of thought and, hence, differ really in but minor ways which may seem important only because they are seen at close range. It holds that mathematics, symbolic logic, philosophy, and so on, are systems contrasted with language which deal directly with this realm of thought, not that they are themselves specialized extensions of language.[3]

This conception has taken such deep root that we are not even aware that it can be subjected to critical analysis. Similarly, we are only aware that we breathe air when we begin to experience a scarcity of it. Whorf gives one more illustration. Suppose that owing to a certain defect in vision a certain people can perceive only the color blue. For them the very term "blue" will be deprived of the meaning which we give it by contrasting it with red, yellow, and the other colors. In the same way, a large majority of people who talk, or at least think, in only one language are simply unaware of the limitations it imposes and the arbitrary element it contains. With nothing with

[3] [Original article, "Science and Linguistics," in *The Technology Review* 42 no 6 (April 1940), Massachusetts Institute of Technology.]

which they can compare their language, its limitations and arbitrary character naturally seem to them universal and unconditional. When linguists conducted critical investigations of large numbers of languages, the structures of which differed greatly, they encountered violations of rules they formerly had considered as universal. It turned out that grammar is not simply an instrument for reproducing thought, but a program and guide for the thinking activity of the individual. Whorf writes:

> "We dissect nature along lines laid down by our native languages. The categories and types that we isolate from the world of phenomena we do not find there because they stare every observer in the face; on the contrary, the world is presented in a kaleidoscopic flux of impressions which has to be organized by our minds—and this means largely by the linguistic systems in our minds." [4]

It should be noted here that Whorf is plainly carried away when he speaks of organizing the stream of impressions, and he incorrectly describes the division of labor between the neuron system and the linguistic system, ascribing the organization of impressions "largely" to the linguistic system. In reality, of course, a very large part of the work of initial organization of impressions is done at the neuron level and what language receives is no longer the raw material, but rather a semifinished product processed in a completely definite manner. Here Whorf makes the same mistake in relation to the neuron system as "natural logic" makes (and Whorf correctly points out!) in relation to the linguistic system. He underestimates the neuron system because it is the same in all people.

It is difficult to conclude that the linguistic system is important for the organization of impressions if we restrict ourselves to a comparison of modern European languages, and possibly also Latin and Ancient Greek. In their fundamental features the systems of these languages coincide, which was an argument in favor of the conception of natural logic. But this coincidence is entirely explained by the fact that the European languages (with minor exceptions) belong to the single family of Indo-European languages, are constructed gener-

[4] "Science and Linguistics."

ally according to the same plan, and have common historical roots. Moreover, for a long period of time they participated in creation of a common culture and in large part this culture, especially in the intellectual area, developed under the determining influence of two Indo-European languages: Greek and Latin. To estimate the breadth of the range of possible grammars one must refer to more linguistic material. The languages of the American Indians, the Hopi, Shawnee, Nutka, and others, serve as such material for Whorf. In comparison with them the European languages are so similar to one another that, for convenience in making comparisons, Whorf consolidates them into one "Standard Average European" language.

■ SUBSTANCE

STANDARD AVERAGE EUROPEAN has two types of nouns which denote material parts of the world around us. Nouns in the first group—such as "a tree," "a stick," "a man," and the like—refer to definite objects which have a definite form. Nouns of the second group—such as "water," "milk," and "meat"—denote homogeneous masses that do not have definite boundaries. There is a very clear grammatical distinction between these groups: the nouns which denote substances do not have a plural case. In English the article before them is dropped, while in French the partitive article is placed in front of them. If we think deeply about the meaning of the difference between these two types of objects, however, it becomes clear that they do not differ from one another so clearly in reality as in language, and possibly there is no actual difference whatsoever. Water, milk, and meat are found in nature only in the form of large or small bodies of definite shape. The difference between the two groups of nouns is thrust upon us by language and often proves so inconvenient that we must use constructions such as "piece of meat" or "glass of water," although the word "piece" does not indicate any definite shape and the word "glass," although it assumes a certain shape, introduces nothing but confusion because when we say "glass of water" we have in mind only a quantity of water, not its shape in the container. Our language would not lose any expressive force if the word "meat" meant a piece of meat and the word "water" meant a certain amount of water.

This is exactly the case in the Hopi language. In their language all nouns denote objects and have singular and plural forms. The nouns we translate as nouns of the second group (substances) do not refer to bodies which have no shape and size, but rather to ones where these characteristics are *not indicated,* where they are ignored in the process of abstraction just as the concept of "stone" does not indicate shape and the concept of "sphere" does not indicate size.

Therefore the concept of substance as something which has material existence and at the same time cannot in principle have any shape could obviously not occur among the Hopi or be understood by a person speaking only the Hopi language. In European culture the concept of substance emerges as a generalization of the concepts which express nouns of the second group while the generalization of concepts which express nouns of the first group leads to the concept of the object. For the Hopi, in whose language there is no division of nouns into two groups, only one generalization is possible and it leads, of course, to the concept of object (or body), for it is possible to abstract from the shape of an observed material object but it cannot be said that it does not exist. The intellectual division of everything existing into a certain nonmaterial form (shape) and a material, but non-form content (substance), which is so typical of traditional European philosophy, will probably seem to the Hopi to be an unnecessary invention. And he will be right! (This is not Whorf's remark, but mine.) The concept of substance, which played such an important part in the arguments among the Medieval Scholastics, has completely disappeared in modern science.

■ THE OBJECTIVIZATION OF TIME

WE WILL NOW take up one more interesting difference between the Hopi language and the Average European Standard. In the European languages the plural forms and cardinal numbers are used in two cases: (1) when they signify an aggregate of objects which form a real group in space and (2) to classify events in time, when the cardinal number does not correspond to any real agregate. We say "ten men" and "ten days." We can picture ten men as a real group, for example ten men on a street corner. But we cannot picture ten days as the aggregate of a group. If it is a group, then it is imagined and consists not of "days," for a day is not an object, but of some objects which are arbitrarily linked to days, for example pages of a calendar or segments in a drawing. In this way we convey a time sequence and a spatial aggregate with the same linguistic apparatus, and it seems to us that this similarity is in the nature of things. In reality this is not true at all. The relations "to be later" and "to be located near" do not have anything in common subjectively. The resemblance between a time sequence and a spatial aggregate is not given to us in perception, but rather in language. This is confirmed by the existence of languages in which there is no such resemblance.

In the Hopi language the plural forms and cardinal numbers are used only to designate objects which may form real groups. The expression "ten days" is not used. Instead of saying "They stayed ten days," the Hopi will say "They left after the tenth day." One cannot say "Ten days is more than nine days," one must say "The tenth day is after the ninth."

Whorf calls the European representation of time *objectivized* because it mentally converts the subjective perception of time as something "which becomes later and later" into some kind of objectively (or, it would be better to say, *objectly*) given objects located in external space. This representation is dictated by our linguistic system, which uses the same numbers both to express temporal relations and to measure spatial quantities and designate spatial relations. This is objectivization. Such terms as "summer," "September," "morning," and "sunset" are nouns in our languages just as the words which designate real objects are. We say "at sunset" just as we say "at a corner," "in September" just as we say "in London."

In the Hopi language all time terms such as summer, morning, and the like are not nouns, they are special adverbial forms (to use the terminology of the Average European Standard). They are a special part of speech which is distinguished from nouns, verbs, and even from other adverbs. They are not used as subjects, objects, or for any other noun function. Of course they have to be translated ''in the summer,'' ''in the morning,'' and so on, but they are not derivatives of any nouns. There is no objectivization of time whatsoever.

In European culture the very concept of ''time'' is a result of the objectivization of the relation of ''earlier-later'' combined with our notion of substance. In our imagination we create nonexistent objects such as ''year,'' ''day,'' and ''second,'' and we call the substance of which they consist ''time.'' We say ''a little time'' and ''a lot of time'' and we ask someone to give us an hour of time as if we were asking for a quart of milk. The Hopi have no basis for a term with this meaning.

The tripartite (past, present, future) verbal system of the Average European Standard directly reflects the objectivization of time. Time is represented as an infinite straight line along which a point is moving (usually from left to right). This point is the present, while to its left is the past and to the right is the future. In the Hopi language, as one might assume, things are different. Their verbs do not have tenses as the European verbs do. Verb forms reflect the source of information and its nature. And this corresponds more closely to reality than the three-tense system. After all, when we say, ''we shall go to the movies tomorrow,'' this does not reflect what will actually occur but only our intention to go to the movies, an intention that exists now and may change at any minute. The same thing applies to past time.

■ LINGUISTIC RELATIVITY

ALL THAT HAS BEEN said in no way leads to the conclusion that the objectivization of time is a bad thing, that we ought to renounce it and change to a Hopi-type language. On the contrary, the most important traits of European culture which have secured such an outstanding place for it—its historical sense (interest in the past, dating, chronicles) and the development of the exact sciences—are linked to the objectivization of time. Science in the only form we yet know it could not have existed without the objectivization of time. The correlation of temporal to spatial relations and the following step, the measurement of time, amounted to the construction of a definite model of sensory experience. It may be that this was the first model created at the level of language. Like any model, it contains an element of arbitrary and willful treatment of reality, but this does not mean that it must be discarded. It must, however, be improved. To improve it, we must conceive of it as a model, not as the primary given. In this respect linguistic analysis is extremely useful because it teaches us to distinguish the relative from the absolute; it teaches us to see the relative and conditional in what at first glance seems absolute and unconditional. Thus, Whorf calls his conception the conception of linguistic relativity.

There is a curious similarity here with the physical theory of relativity. Objectivized time is the foundation of classical Newtonian mechanics. Because the imagined space into which we project time is in no way linked to real space, we picture time as something that "flows" evenly at all points in real space. Einstein dared to reconsider this notion and showed that it is not upheld in experimental data and that it should be rejected. But as we know very well, this rejection does not come without difficulty, because, as Whorf writes: "The offhand answer, laying the blame upon intuition for our slowness in discovering mysteries of the cosmos, such as relativity, is the wrong one. The right answer is: Newtonian space, time, and matter are no intuitions. They are recepts from culture and language. That is where Newton got them." [5] Once again here we should temper the statements of the enthusiastic linguist. Newtonian concepts, of course, rely directly on our intuition. But this intuition itself is not a pure reflection of primary sensory experience, of the "kaleioscopic flux of impressions"; rather it is a product of the organization of this experience, and language and culture really do play a considerable part in this organization.

■ THE METASYSTEM TRANSITION IN LANGUAGE

LANGUAGE EMERGES when the phenomena of reality are encoded in linguistic objects. But after its origin language itself becomes a phe-

[5] [Original article, "The Relation of Habitual Thought and Behavior to Language," published in *Language, Culture, and Personality* (Menasha, Wisconsin: Sapir Memorial Publication Fund 1941), pp. 75–93.]

nomenon of reality. Linguistic objects become very important elements of social activity and are included in human life like tools and household accessories. And just as the human being creates new tools for the manufacture and refinement of other tools so he creates new linguistic objects to describe the reality which already contains linguistic objects. A metasystem transition within the system of language occurs. Because the new linguistic objects are in their turn elements of reality and may become objects of encoding, the metasystem transition may be repeated an unlimited number of times. Like other cybernetic systems we have considered in this book, language is a part of the developing universe and is developing itself. And like other systems, language—and together with it thinking—is undergoing qualitative changes through metasystem transitions of varying scale, that is to say, transitions which encompass more or less important subsystems of the language system.

With all the physical-chemical differences that exist between the linguistic system and the neuronal system it is easy to see that, functionally, metasystem transitions in language are a natural continuation of the metasystem transitions in the neuronal structures, serving to create more highly refined models of reality. To clarify this thought let us look again at the diagram in figure 7.4, this time viewing it as a diagram of a device for processing information coming from an illuminated screen and, consequently, as a partial (and crude) model of the organization of the brain. In the diagram we see classifiers which correspond to the concepts of "spot," "contour," "inside," and "enter into." These concepts stand at different levels of the hierarchy and the number of levels is in principle unlimited. But let us ask: is it possible that there could be a metasystem transition of such large scale that it would be represented not by adding a new level to figure 7.4 but as a departure from the plane of the drawing in general, as the creation of a new plane?

If we compare our artificial system to real biological systems it corresponds to a nerve net with a rigidly fixed hierarchy of concepts. This is the stage of the complex reflex. To reach a new plane would signify the transition to the stage of associating, when the system of connections among classifiers becomes controlled.

The concepts involved in figure 7.4 are taken from language. In addition, there are in language concepts that "go outside the plane" of the diagram. Regarding the concept "inside" we can say that it is an example of a spatial relation among objects. Other examples of spatial relations are the concepts "touches," "intersects," and "between." Classifiers to recognize these concepts could be added to the diagram. But how about the very concept "spatial relation"? It is the sought-for metaconcept in relation to the concepts "inside," "between," and so on; its relation to them is that of name to meaning. If we were able to think of a way to embody the concept of "spatial relation" in the form of some kind of device that supplements the device in figure 7.4, it would plainly have to form a metasystem in relation to such classifiers as "inside," "between," and others. The task it would be able to perform would be modifying the structure of work of these classifiers or creating new ones that recognize some new spatial relation. But is not the very purpose of the appearance of the concept of "spatial relation" in language itself to achieve a better understanding of how the linguistic system works—to modify it and create new concepts? Most certainly it is. The metasystem transition in the development of language performs the same role as it does in the development of neuronal structures.

■ THE CONCEPT-CONSTRUCT

CONCEPTS SUCH AS that of "spatial relation" rely on reality indirectly, through the mediation of intermediate linguistic structures. They become possible as a result of a certain linguistic construction, and therefore we shall call them *constructs*. Statements containing constructs demand a certain linguistic activity to establish their truth or falsehood. Concept-constructs do not exist outside the linguistic system. For example, the concept of "spatial relation" cannot arise where there are no words "inside," "between," and so on, although the corresponding neuronal concepts may have existed for a long time.

We can now make a survey of the levels of language viewed as a control hierarchy. We shall take the signals of animals for the zero level of language. The appearance of the standard actions of affirmation and negation, logical connectives, and predicates is, as we have already said, a metasystem transition. They create the first level of language. The next metasystem transition forms the second level of language, whose concepts are constructs. Among the concepts are grammar and logic. At the first level, grammar and logic are the highest control systems that create language but are not themselves subject to control; however, at the second level they become objects of study and control (artificial construction). The second level of language may be called the level of constructs, and also the level of self-description.

The level of development of language determines the relation between the linguistic and the neuronal systems. At the zero level, language transmits only elementary control information; at the first level it acquires the ability to fix and transmit certain models of reality, but only those models which already exist at the neuronal level. First-level language may be represented as a copy or photograph of neuronal models (taking into account the inverse of language as a corrective). Finally, at the level of constructs language becomes able to fix models of reality which could not (bearing in mind the given biological species of the human being) occur at the neuronal level. Such models are called *theories*.

We have cited numbers and operations with them as the simplest and most graphic example of models that do not exist at the neuronal level and are created at the language level. Arithmetic was one of the first theories created by the human race. It is easy to see that numbers, or more precisely large numbers, are constructs. Neuronal concepts correspond to the numbers two and three; we distinguish two objects from three and from one at the first glance. But the number 137 is a construct; it has meaning only to the extent that the number 136 has meaning, which in its turn relies on the number 135 and so on. Here there is a metasystem transition, the emergence of the process of counting which generates concrete numbers. Within the framework of the metasystem of counting, a hierarchy by complexity arises: the natural series of numbers. The appearance of the concept of "number" marks a new metasystem transition which assumes that counting has become an established part of everyday life. An abstract concept of "number" is not required for counting; it only becomes necessary when people begin to think about counting. The concept of the number is a construct of a higher level than concrete numbers. The concepts of arithmetic operations are located at the same level.

On the second level of language we have consolidated all the concepts which do not rely directly on neuronal concepts but rather require auxiliary linguistic constructions. With such a definition the second level is the last one formally, but it contains a control hierarchy that forms through metasystem transitions and may in principle be as high as one likes. We have seen this in the example of concrete numbers and the concept of "number." Metasystem transitions can differ in scale and occur in relation to different subsystems of language. Therefore, second-level language has a complex structure which can be figuratively pictured not in the form of even layers lying one upon the other but in the form of a building or complex of buildings with vertical and horizontal structure. Different control hierarchies and hierarchies of complexity generated by the subsystems become interwoven and form a multifaceted architectural complex. Second-level language is the language of philosophy and science. First-level language is ordinarily called "everyday" or "conversational" language.

■ THE THINKING OF HUMANS AND ANIMALS

IT IS SOMETIMES SAID that the human being can think in abstract concepts, whereas abstract concepts are inaccessible to the animal, who can attain only a few concrete concepts. If the term "abstract" is understood (as is the case here) to mean devoid of nonessential characteristics, this assertion will not withstand even the slighest criticism. We have seen that the crucial distinguishing feature in human thinking is the presence of control of associations, which manifests itself above all as a capability for imagination. As for a difference in the concepts, in any case it cannot be reduced to an opposition between abstract and concrete. Every concept is abstract. The concept of cat is abstract for the dog because, for example, it contains an abstraction from the coloring of the cat (a nonessential characteristic). If we measure mental capabilities by the degree of abstraction of concepts the frog will prove to be one of the most intelligent animals, for it thinks with just two concepts, albeit extremely abstract ones: "something small and rapidly moving" and "something large, dark, and not moving very rapidly." As you see, our language does not even have special terms for these concepts.

The truly profound difference between the conceptual apparatus of higher animals and that of human beings is that animals cannot attain concept-constructs; these concepts assume a capability for linguistic activity. It is not abstract concepts which distinguish human thinking; it is concept-constructs. In partial justification of the statement above, we should note that the expression "abstract concept" is commonly used to refer to precisely what we call the concept-construct, and people talk about the degree of abstraction where they should actually speak of the "construct quality" ["konstrukt-nost' "—the degree to which constructs are used—trans.]. It is true that the concept of number is formed by abstraction starting from concrete numbers and that the concept of the spatial relation begins from concrete relations; but the distinctive feature here is not the actual process of abstraction (which, as we have seen, appeared in the very early stages of the cybernetic period of life), rather it is the fact that in the process of abstraction linguistic objects play the most essential part. The principal thing here is *construction,* not abstraction. Abstraction without construction simply leads to loss of meaning, to concepts such as "something" and "some."

CHAPTER EIGHT
Primitive Thinking

■ THE SYSTEM ASPECT OF CULTURE

LET US CONTINUE our excursion through the stages of evolution. The subject of our analysis now will be the history of the development of language and thinking, the most important component of "spiritual" culture. As we have already noted, the division of culture into "material" and "spiritual" is quite arbitrary and the terms themselves do not reflect the substance of the division very accurately, so that when we want to emphasize this we place them within quotation marks. The use of a tool and, even more so, the creation of new ones demand the work of imagination and are accompanied by emotions, giving us grounds to consider these phenomena part of "spiritual" culture. At the same time, the process of thinking manifests itself as definite linguistic activity directed to completely material objects— linguistic objects. Language and thinking are very closely interconnected with material culture. The historian who sets himself the task of investigating the mechanism of the development of culture can only consider these phenomena in their interrelationship. He must also take account of other aspects of culture—above all the social structure of society—as well as the influence of natural conditions, historical accidents, and other factors. But the present investigation is not historical. Our task is simpler: without going into the details of historical development to describe what happened from a cybernetic or, as is also said, from a *systems* point of view. As with the question of the origin of human beings, we shall not be interested in a profound, intricate presentation of the historical circumstances that led to the particular step in the development of culture at the particular place and time. Our approach remains very global and general. We are interested in just one aspect of culture (but it is the most important one in the mechanics of development!)—its structure as a control hierarchy. Accordingly, we will view the development of culture also as a process of increasing complexity in this hierarchy through successive metasystem transitions. We will show, as was also true in the case of biological development, that the most important stages in the development of language and thinking are separated from one another by precisely these metasystem transitions.

■ THE SAVAGE STATE AND CIVILIZATION

IN THE DEVELOPMENT of culture we discern above all two clearly distinct steps: the *savage state* (primitive culture) and *civilization*. The clear delineation between them does not mean that there are no transitional forms at all; the transition from the savage state to civilization is not carried out instantaneously, of course. But once it has begun, the development of culture through the creation of civilization takes place so rapidly that an obvious and indisputable difference between the new level of culture and the old manifests itself in a period of time which is vastly smaller than the time of existence in the savage state. The emergence of civilization is a qualitative leap forward. The total time of existence of civilization on Earth (not more than 5,000 to 6,000 years) constitutes a small part of the time (at least 40,000 years) during which the human race has existed as a biologically invariable species. Thus, the emergence of civilization is a phenomenon which belongs entirely to the sphere of culture and is in no way linked to the biological refinement of the human being. This distinguishes it from the emergence of language and labor activity but the consequences of this phenomenon for the biosphere are truly enormous, even if they are measured by simple quantitative indexes rather than by the complexity of the structures which emerge. In the short time during which civilization has existed, the human race has had incomparably more effect on the face of the planet than during the many millennia of the savage state. The size of the human race and its effect on the biosphere have grown at a particularly swift pace in the last three centuries; this is a result of the advances of science, the favorite child of civilization.

This fact requires explanation. Such an abrupt qualitative leap forward in the observed manifestations of culture must be linked to some essential, fundamental change in the internal structure of culture. Language is the core of culture; it insures its uniformity, its "nervous system." We have in mind here not language as an abstract system possessing particular grammatical characteristics and used for expressing thoughts, but rather language as a living reality, as the social norm of linguistic activity. In other words, we have in mind the full observed (material if you like) side of thinking. Therefore, when we say "language" we immediately add "and thinking." So language (and thinking) are the nervous system of culture and it may therefore be expected that there is some important difference between the language and thinking of primitive and of modern peoples. Indeed, a study of the culture of backward peoples reveals that they have a way of thinking which greatly differs from that of modern Europeans. This difference is by no means simply one in levels of knowledge. If a European is placed under primitive conditions he will hardly be able to use (or even show!) his knowledge of Ohm's law, the chemical formula for water, or the fact that the Earth revolves around the Sun and not vice versa. But the difference in way of thinking, in the approach to the phenomena of reality, remains and will quickly show itself in behavior.

That difference can be summarized as follows. To a primitive person the observed phenomena of the world appear to be caused by invisible, supernatural beings. The primitives resort to incantations, ritual dances, sacrifices, strictly observed prohibitions (tabus), and so on to appease or drive off such beings. E. Taylor, one of the founders of the scientific study of primitive cultures, has called this view of the world *animism,* assuming the existence of spirits in all objects. To primitive people, certain mysterious relations and influences can exist between different objects ("mystic participation," in the terminology of the French ethnographer L. Lévy Bruhl). Such relations always exist, in particular, between the object and its image or name. From this follow primitive magic and belief in the mystical connection between the tribe and a particular animal species (the totem).

But what is most surprising to the European is not the content of the representations of primitive people, rather it is their extreme resistance and insensitivity to the data of experience. Primitive thinking is inconceivably conservative and closed. Obvious facts which, in the European's opinion, would inevitably have to change the notions of the primitive individual and force him to reconsider certain convictions do not, for some reason, have any effect on him at all. And attempts to persuade and prove often lead to results diametrically opposite to what was expected. It is this, not the belief in the existence of spirits and a mystical connection among objects, which is the most profound difference between primitive and modern thinking. In the last analysis, everything in the world is truly interrelated! When presenting the law of universal gravity we could say that there is a spirit of gravity in every body and each spirit strives to draw closer to the other spirit with a force proportional to the mass of the two bodies and inversely proportional to the square of the distance between them. This would not hinder us at all in correctly calculating the movement of the planets. But even if we do not use the word "spirit," we still use the word "force." And, in actuality, what is the Newtonian force of gravity? It is the same spirit: something unseen, unheard, unfelt, without taste or smell, but nonetheless really existing and influencing things.

These characteristics of the thinking of primitive people are amazingly widespread. It can be said that they are common to all primitive peoples, regardless of their racial affiliation and geographic conditions and despite differences in the concrete forms of culture where they manifest themselves. This gives us grounds to speak of *primitive thinking,* juxtaposing it to modern thinking and viewing it as the first, historically inevitable phase of human thinking. Without negating the correctness of such a division or of our attempts to explain the transition, it should be noted that, as with any division of a continuous process into distinct phases, there are transitional forms too; in the thinking of a modern civilized person we often discern characteristics that go back to the intellectual activity of mammoth- and cave-bear hunters.

■ THE METASYSTEM TRANSITION IN LINGUISTIC ACTIVITY

THE PRIMITIVE PHASE is the phase of thinking which follows immediately after the emergence of language and is characterized by the fact that linguistic activity has not yet become its own object. The transition to the phase of modern thinking is a metasystem transition, in which there is an emergence of linguistic activity directed to linguistic activity. The language of primitive people is first-level language, while the language of modern people is second-level language (which specifically includes grammar and logic). But the transition to modern thinking is not simply a metasystem transition in language if we view language statically, as a certain possibility or method of activity. It includes a metasystem transition in real linguistic activity as a socially significant norm of behavior. With the transition to the phase of modern thinking it is not enough to think about something; one must also ask why one thinks that way, whether there is an alternative line of thought, and what would be the consequences of these particular thoughts. Thus, modern thinking is critical thinking, while primitive thinking can be called precritical. Critical thinking has become so accepted that it is taken for granted today. It is true that we sometimes say that a particular individual thinks "uncritically"; however, the term itself means that uncritical thinking is the exception, not the rule. An uncritical quality in thinking is ordinarily considered a weakness, and attempts are made to explain it in some way— perhaps by the influence of emotions, a desire to avoid certain conclusions, and so on. In the case of certain convictions (dogmas, for example), uncritical thinking may be justified by their special (or sacred) origin. But the general stream of our thinking continues to be critical. This does not mean that it is always original and free of stereotypes, but even when we think in stereotyped ways we are nonetheless thinking critically because of the nature of the stereotype. It includes linguistic activity directed to linguistic activity, it teaches us to separate the name from the meaning and remember the arbitrary nature of the connection between them, and it teaches us to think, "Why do I talk or think this way?" Not only do we use this stereotype, we also employ the results of its use by preceding generations.

Things are different in primitive society, where the relation between language and reality is not yet the object of thought. There the social norm of thinking is to treat the words, notions, and rules of one's culture as something unconditionally given, absolute, and inseparable from other elements of reality. This is a very fundamental difference from the modern way of thinking. Let us consider primitive thinking in more detail and show that its basic observed characteristics follow from this feature, its *precritical nature*.

We use below material from the writings of L. Lévy-Bruhl (in Russian, *Pervobytnoe myshlenie* [Primitive Thinking], Ateist Publishing House, 1930; this book combines material from Lévy-Bruhl's *La mentalité primitive* and *Les fonctions mentales dans les sociéetés inférieures*). This book is interesting because it collects a great deal of material on primitive culture which convincingly demonstrates the difference between primitive and modern thinking. A feature of Lévy-Bruhl's conception is that he describes the thinking of individual members of primitive society as controlled by the *collective representations* of the given culture (actually, of course, this does not apply only to primitive society, but Lévy-Bruhl somehow does not notice this). Also to Lévy-Bruhl's credit is his observation that collective representations in primitive society differ fundamentally from our own and therefore it is completely incorrect to explain the thinking of a primitive person by assuming (often unconsciously) that he is modern. The rest of Lévy-Bruhl's conception is quite unimportant. He describes primitive thinking as "prelogical," "mystically oriented," and "controlled by the law of participation." These concepts remain very vague and add nothing to the material which has been collected. Only the term "prelogical" thinking arouses our interest; it resembles our definition of primitive thinking as precritical.

■ THE MAGIC OF WORDS

THE ASSOCIATION name-meaning L_i-R_i already exists in primitive thinking, for language has become a firmly established part of life; but the association has not yet become an object of attention, because the metasystem transition to the second level of linguistic activity still has not taken place. Therefore the association L_i-R_i is perceived in exactly the same way as any association R_i-R_j among elements of reality, for example the association between lightning and thunder. For primitive thinking the relation between an object and its name is an absolute (so to speak physical) reality which simply cannot be doubted. In fact—and this follows from the fundamental characteristic of the association—the primitive person thinks that there is a single object L_iR_i whose name L_i and material appearance R_i are different parts or aspects. Many investigators testify to the existence of this attitude toward names among primitive peoples. "The Indian regards his name, not as a mere label, but as a distinct part of his personality, just as much as are his eyes or his teeth, and believes that injury will result as surely from the malicious handling of his name as from a wound inflicted on any part of his physical organism. This belief was found among various tribes from the Atlantic to the Pacific." [1] Therefore many peoples follow the custom of not using a person's "real" name in everyday life, but instead using a nickname which is viewed as accidental and arbitrary. A. B. Ellis, who studied the peoples of West Africa, states that they "believe that there is a real and material connection between a man and his name, and that by means of the name injury may be done to the man. . . . In consequence of this belief the name of the king of Dahomi is always kept secret. . . . It appears strange that the birth-name only, and not an alias, should be believed capable of carrying some of the personality of the bearer elsewhere . . . but the native view seems to be that the alias does not really belong to the man." [2] This division of names

---

[1] Original in James Mooney, "The Sacred Formulas of the Cherokee," 7th Annual Report of the Bureau of Ethnology, Washington, GPO, 1885–1886, p. 343—trans.
[2] Original A. B. Ellis, *The Ewe-Speaking People,* London, 1890, p. 98—trans.

into "real" and "not real" is obviously the first step on the path toward the metasystem transition.

The relation between an object and its image is perceived in exactly the same way as between an object and its name. In general, primitive thinking does not make any essential distinction between the image and the name. This is not surprising, because the image is connected with the original of the same association that the name is. The image is the name and the name is the image. All images and names of an object taken together with the object itself form a single whole something (specifically a representation created by an association). Therefore it seems obvious that when we act on a part we act by the same token on the whole, which also means on its other parts. By making an image of a buffalo pierced by an arrow the primitive believes that he is fostering a successful hunt for a real buffalo. G. Catlin, an artist and scientist who lived among the Mandans of North America notes that they believed the pictures in the portraits he made borrowed a certain part of the life principle from their original. One of the Mandans told him that he knew he had put many buffalo in his book because the Indian was there while he drew them and after that observed that there were not so many buffalo for food. Obviously the Indian understood that the white man was not literally putting buffalo in his book; but it was nevertheless obvious to him that in some sense (specifically in relation to the real-buffalo–buffalo-picture complexes) the white man was putting the buffalo in his book, because their numbers declined. The word "put" [the Russian *ulozhit'*—to put in, pack, fit] is used here in a somewhat metaphorical sense if the primary meaning refers to an action on a "material" buffalo, but this does not affect the validity of the thought. Many terms in all the world's languages are used metaphorically, and without this the development of language would be impossible. When we use the Russian expression *ulozhit' sya v golove* [literally—to be packed, fit in the head; the idiomatic meaning is "to be understood"] we do not mean that something has been put in our head in the same way that items are packed in a suitcase.

■ SPIRITS AND THE LIKE

NOW LET US MOVE ON to "spirits," which play such an important part in primitive thinking. We shall see that the appearance of supernatural beings is an inevitable consequence of the emergence of language and that they disappear (with the same inevitability as they appeared) only upon the metasystem transition to the level of critical thinking.

First let us think about the situation where language already exists but its relation to reality still has not become an object of study. Thanks to language, something like a doubling of objects occurs: instead of object R_i a person deals with a complex R_iL_i where L_i is the name of R_i. In this complex, the linguistic object L_i is the more accessible and, in this sense, more *permanent* component. One can say the word "sun" regardless of whether the sun is visible at the particular moment or not. One can repeat the name of a person as often as one likes while the person himself may be long dead. Each time his face will rise up in the imagination of the speaker. As a result the relation between the name and the meaning becomes inverted: the object L_i acquires the characteristics of something primary and the object R_i becomes secondary. The normal relation is restored only after the metasystem transition, when R_i and L_i are equally objects of attention, and the connection between them is of special importance. Until this has happened the word L_i plays the leading role in the complex R_iL_i, and the faithful imagination is ready to link any pictures with each word used in social linguistic practices. Some words of the language of primitive culture signify objects which really exist from our modern point of view while others signify things which from our point of view do not really exist (spirits and so on). But from the point of view of the primitive individual there is no difference between them—or perhaps simply a quantitative one. Ordinary objects may or may not be visible (perhaps they are hidden; perhaps it is dark). They may be visible only to some. The same is true of spirits, only it is harder to see them; either no one sees them or they are seen by sorcerers. Among the Klamath Indians in North America the medicine man who was summoned to a sick person had to consult with the spirits of certain animals. Only one who had gone through a five-year course of preparation to be a medicine man could see these spirits, but he saw them just as plainly as the objects around himself. The Taragumars believed that large snakes with horns and enormous eyes lived in the rivers. But only shamans could see them. Among the Buryats the opinion was widespread that when a child became dangerously ill the cause was a little animal called an *anokkha,* which was eating the top of the child's head away. The *anokkha* resembled a mole or cat, but only shamans could see it. Among the Guichols there is a ritual ceremony in which the heads of does are placed next to the heads of stags and it is considered that both the does and the stags have antlers, although no one except the shamans see them.

There is an enormously broad variety of invisible objects in the representations of primitive peoples. They are not just formless spirits, but also objects or beings which have completely defined external appearances (except that they are not always perceived and not perceived by all). Language provides an abundance of material for the creation of imagined essences. Any quality is easily and without difficulty converted into an essence. The difference between a living person and a dead one produces the soul, and the difference between a sick person and a healthy one gives us illness. The representation of illness as something substantial, objective, which may enter and depart from a body and move in space, is perhaps typical of all primitive peoples. The same thing is true of the soul. It is curious that just as there are different illnesses, among some peoples there also exist different ''souls'' in the human being. According to the observations of A. B. Ellis the Negroes of the West African coast distinguish two human spirits: *kra* and *sraman.* Kra lives in the person as long as he is alive but departs when the person sleeps; dreams are the adventures of the kra. When a person dies his kra may move to the body of another person or animal, but it may instead wander the world. The sraman forms only upon the death of the person and in the land of the dead continues the way of life which the deceased had followed.

This belief shows even more clearly among the American Indians. The Mandans, for example, believe that every person carries several spirits: one of them is white, another is swarthy, and the third is a light color. The Dakotas believe that a person has four souls: the corporal soul, which dies along with the person; the spirit, which lives with the body or near it; the soul responsible for the actions of the body; and the soul that always remains near a lock of the deceased's hair, which is preserved by relatives until it can be thrown onto enemy territory, whereupon it becomes a wandering ghost carrying illness and death. G. H. Jones, a scientist who studied beliefs in Korea, writes of spirits that occupy the sky and everywhere on earth. They supposedly lie in wait for a person along the roads, in the trees, in the mountains and valleys, and in the rivers and streams. They follow the person constantly—even to his own home, where they have settled within the walls, hang from the beams, and attach themselves to the room dividers.

■ THE TRASH HEAP OF REPRESENTATIONS

AS WE HAVE NOTED, it is not the fact of belief in the existence of invisible things and influences that distinguishes primitive thinking from modern thinking, but the content of the representations and particularly the relation between the content and the data of experience. We believe in the existence of neutrons although no one has ever seen them and never will. But we know that all the words in our vocabulary have meaning only to the extent that, taken together, they successfully describe observed phenomena and help to predict them. As soon as they stop fulfilling this role, as a result of new data from experience or owing to reorganization of the system of word use (theory), we toss them aside without regret. That is what happened, for example, with "phlogiston" or ether. Even earlier, all kinds of imagined beings and objects which were so typical of the thinking of our ancestors disappeared from language and thinking. What irritates us in primitive thinking is not the assumption of the existence of spirits but rather that this assumption, coming together with certain assumptions about the traits and habits of the spirits, explains nothing at all and often simply contradicts experience. We shall cite a few typical observations by investigators. In his Nicobar Island diaries,

V. Solomon wrote: "The people in all villages have performed the ceremony called 'tanangla,' signifying either 'support' or 'prevention.' This is to prevent the illness caused by the north-east monsoon. Poor Nicobarese! They do the same thing year after year, but to no effect." [3]

And M. Dobrizhoffer observed that

A wound inflicted with a spear often gapes so wide that it affords ample room for life to go out and death to come in; yet if the man dies of the wound they madly believe him killed not by a weapon but by the deadly arts of the jugglers. . . . They are persuaded that the juggler will be banished from amongst the living and made to atone for their relation's death if the heart and tongue be pulled out of the dead man's body immediately after his decease, roasted at the fire and given to dogs to devour. Though so many hearts and tongues are devoured, and they never observed any of the jugglers die, yet they still religiously adhere to the custom of their ancestors by cutting out the hearts and tongues of infants and adults of both sexes, as soon as they have expired. [4]

Because primitive people are unable to make their representations an object of analysis, these representations form a kind of trash heap. The trash heap accumulates easily but no one works to clean it up. For the primitive there are not and cannot be meaningless words. If he does not understand a word it frightens him as an unfamiliar animal, weapon, or natural phenomenon would. An opinion which has arisen as a result of the chance combination of circumstances is preserved from generation to generation without any real basis. The explanation of some phenomenon may be completely arbitrary and nonetheless fully satisfy the primitive. Critical thinking considers each explanation (linguistic model of reality) alongside other, competing explanations (models) and it is not satisfied until it is shown that the particular explanation is better than its rivals. In logic this is called the law of sufficient grounds. The law of sufficient grounds is

[3] Original in V. Solomon, "Extracts from Diaries Kept in Car Nicobar," *Journal of the Anthropological Institute of Great Britain and Ireland* 32 (January–June 1902): 213—trans.
[4] Original in M. Dobrizhoffer, *An Account of the Abipones,* (London, 1822), vol 2, p 223—trans.

absolutely foreign to precritical thinking. It is here that the metasystem transition which separates modern thinking from primitive thinking is seen most clearly.

Thanks to this characteristic the primitive's belief in the effectiveness of magic, incantations, sorcery, and the like is unconquerable. His "theory" gives an explanation (often not just one but several!) for everything that happens around him. He cannot yet evaluate his theory—or even individual parts of it—critically. P. Bowdich tells of a savage who took up a fetish which was supposed to make him invulnerable. He decided to test it and let himself be shot in the arm; it broke his bone. The sorcerer explained that the offended fetish had just revealed to him the cause of what had happened: the young man had had sexual relations with his wife on a forbidden day. Everyone was satisfied. The wounded man admitted that it was true and his fellow tribesmen were only reinforced in their belief. Innumerable similar examples could be given.[5]

■ BELIEF AND KNOWLEDGE

WHEN WE SAY that a primitive person believes in the existence of spirits or certain actions by them we predispose ourselves to an incorrect understanding of his psychology. When speaking of belief we juxtapose it to knowledge. But the very difference between belief and knowledge emerges only at the level of critical thinking and reflects a difference in the psychological validity of representations, which follows from the difference in their sources. For a primitive there is no difference between belief and knowledge and his attitude toward his representations resembles our attitude toward our knowledge, not our beliefs. From a psychological point of view the primitive person *knows* that spirits exist, he knows that incantations can drive out illness or inflict it, and he knows that after death he will live in the land of the dead. Therefore we shall avoid calling the primitive person's worldview primitive religion; the terms "primitive philosophy"

[5] T. Edward Bowdich, *Mission from Cape Coast Castle to Ashantee,* (London: Frank Cass and Co., Ltd, 3rd ed, 1966), p 439—trans.

or "primitive science" have equal right to exist. These forms of activity can only be distinguished at the level of critical thinking. This refers both to the difference between belief and knowledge and to the difference between the "otherworldly" and that which is "of this world." The fact that the representations of primitive people involve spirits, ghosts, shadows of the dead, and other devil figures still does not make these representations religious, because all of these things are perceived as entirely of this world and just as real (material if you like) as the animals, wind, or sunlight. L. Lévy-Bruhl, who defines the psychological activity of primitive man as mystic, nonetheless emphasizes that this is not at all the same as mysticism in the modern meaning of the word. "For lack of a better term," he writes, "I am going to use this one; this is not because of its connection with the religious mysticism of our societies, which is something quite different, but because in the narrowest meaning of the word 'mystic' is close to belief in forces, influences, and actions which are unnoticed and intangible to the senses but real all the same." Many observers are struck by how real the shadows or spirits of their ancestors seem to primitive peoples. R. Codrington writes about the Melanesians: [6] When a native says that he is a person, he wants it understood that he is a person not a *spirit*. He does not mean that he is a person not an *animal*. To him, intelligent beings in the world are divided into two categories: people who are alive and people who have died. In the Motu tribe this is *ta-mur* and *ta-mate*. When the Melanesians see white people for the first time they take them for *ta-mate,* that is, for spirits who have returned to life, and when the whites ask the natives who they are, the latter call themselves *ta-mur,* that is, people not spirits. Among the Chiriguanos of South America when two people meet they exchange this greeting: "Are you alive?"—"Yes, I am alive." Some other South American tribes also use this form.

■ THE CONSERVATISM OF PRECRITICAL THINKING

CONSERVATISM is inherent in precritical thinking; it is a direct consequence of the absence of an apparatus for changing linguistic models.

[6] *The Melanesian Languages* (Oxford, 1891)

All conceivable kinds of rules and prohibitions guide behavior and thinking along a strictly defined path sanctified by tradition. Violation of traditions evokes superstitious terror. There have been cases where people who accidentally violated a tabu died when they learned what they had done. They knew that they were supposed to die and they died as a result of self-suggestion.

Of course, this does not mean that there is no progress whatsoever in primitive society. Within the limits of what is permitted by custom, primitive people sometimes demonstrate amazing feats of art, dexterity, patience, and persistence. Within the same framework tools and weapons are refined from generation to generation and experience is accumulated. The trouble is that these limitations are extremely narrow and rigid. Only exceptional circumstances can force a tribe (most likely the remnants of a tribe which has been destroyed by enemies or is dying from hunger) to violate custom. It was probably in precisely such situations that the major advances in primitive culture were made. A people which has fallen into isolation and owing to unfavorable natural conditions is not able to multiply and break up into bitterly hostile peoples may maintain its level of primitive culture unchanged for millennia.

In the stage of precritical thinking, language plays a paradoxical role. In performance of its communicative function (communication among people, passing experience down from generation to generation, stablizing social groups) it is useful to people. But then its noncommunicative, modeling function causes more harm than good. This refers to those models which are created not at the level of the association of nonlinguistic representations but only at the level of language, that is, primarily the primitive "theory of spirits." As we have already noted, the communicative function itself becomes possible only thanks to the modeling function. But as long as linguistic models merely reflect neuronal models we speak of the purely communicative functions; when new models (theories) are created we speak of the noncommunicative function. In primitive society we see two theories: the rudiments of arithmetic (counting by means of fingers, chips, and the like) and the "theory of spirits." Arithmetic is, of course, a positive phenomenon, but it does not play a major part in primitive life and is in fact absent among many peoples; the "theory of spirits," on the contrary, permeates all primitive life and has a negative influence on it. And this is the paradox. The first independent steps of the linguistic system, which should according to the idea lead to (and later in fact do lead to) an enormous leap forward in modeling reality, at first produce poisonous discharges which retard further development. This is a result of the *savage,* so to speak, growth of the "theory of spirits." It can be compared with a weed which sprouts on well-fertilized soil if the garden is not managed. As we have seen, the weed's seeds are contained in the soil itself, in language. Only the transition to the level of critical thinking (careful cultivation of the soil, selection of plants for crops, and weed control) produces the expected yield.

■ THE EMERGENCE OF CIVILIZATION

WE KNOW THAT this transition took place. The emergence of critical thinking was the most important milepost of evolution after the appearance of the human being. Critical thinking and civilization arise at the same time and develop in close interdependence. Increasing labor productivity, contacts among different tribal cultures, and the breakup of society into classes all inexorably weaken traditional tribal thinking and force people to reflect upon the content of their representations and compare them with those of other cultures. In this way critical thinking takes root and gradually becomes the norm. On the other hand, critical thinking emancipates people and leads to a sharp rise in labor productivity and to the appearance of new forms of behavior. Both processes support and reinforce one another; society begins to develop swiftly. There is a kind of 180 degree turn in the vector of society's interest: in primitive society it is directed backward, to the past, to observance of the laws of ancestors; in a developing situation, at least among part of society (the "creative minority" according to A. Toynbee), it is pointed forward, into the future, toward change in the existing situation. Thanks to a metasystem transition culture acquires dynamism and its own internal impetus toward development. The redirection of language activity to itself creates the stairway effect: each level of logical (language) thinking, which has emerged as a result of the analysis of logical thinking, becomes, in its turn, an object of logical analysis. Critical thinking is an ultrametasystem capable of self-development. Primitive tribal cultures evolve by the formation of groups and the struggle for existence among them, just as in the animal world. Civilization evolves under the influence of internal factors. It is true that the civilizations of the past typically stopped in their development upon reaching a certain level, but all the same the leaps forward were extremely great in comparison with the advances of primitive cultures, and they grew larger as critical thinking became ever more established. Modern civilization is global, so that the factor of its struggle for existence as a whole (that is to say, against rivals) disappears and all its development occurs exclusively through the action of internal contradictions. Essentially, it was only with the transition to the level of critical thinking that the revolutionary essence of the emergence of thinking manifested itself and the age of reason began in earnest.

In the process of a metasystem transition there is, as we know, a moment when the new attribute demonstrates its superiority in a way which cannot be doubted, and from this moment the metasystem transition may be considered finally and irreversibly completed. In the transition to critical thinking this moment was the culture of Ancient Greece, which it is absolutely correct to call the cradle of modern civilization and culture. At that time, about 2,500 years ago, philosophy, logic, and mathematics (mathematics in the full sense of the word, that is to say, including proof) emerged. And from that time critical thinking became the recognized and essential basis of developing culture.

CHAPTER NINE
Mathematics before the Greeks

■ NATURE'S MISTAKE

WE HAVE ALREADY mentioned the process of counting as an example of using a model of reality that is not contained in the brain but is created at the level of language. This is a very clear example. Counting is based on the ability to divide the surrounding world up into distinct objects. This ability emerged quite far back in the course of evolution; the higher vertebrates appear to have it in the same degree as humans do. It is plain that a living being capable of distinguishing separate objects would find it useful in the struggle for existence if it could also count them (for example, this would help one become oriented in an unfamiliar area). Description by means of numbers is a natural, integrated complement to differential description by recognition of distinct objects. Yet the cybernetic apparatus for recognizing numbers, for counting, can be extremely simple. This task is much easier than distinguishing among separate objects. Therefore one would expect that, within limits imposed by the organization of the organs of sight, recognition of numbers would have appeared in the course of evolution. The human eye can distinguish tens and hundreds of distinct objects at once. We might expect that human beings would be able to tell a group of 200 objects from a group of 201 just as easily as we tell two objects from three.

But nature did not wish or was unable to give us this capability. The numbers which are immediately recognizable are ridiculously few, usually four or five. Through training certain progress can be made, but this is done by mentally breaking up into groups or by memorizing pictures as whole units and then counting them in the mind. The limitation on direct discrimination remains. It is in no way related to the organization of the organs of sight and apparently results from some more deep-seated characteristics of brain structure. We do not yet know what they are. One fact forces us to ponder and suggests some hypotheses:

In addition to spatial discrimination of numbers there is temporal discrimination. You never confuse two knocks at a door with three or one. But eight or ten knocks is already, no doubt, "many" and we can only distinguish such sounds by their total length (this corresponds to the total area occupied by homogeneous objects in spatial discrimination). The limit which restricts both types of discrimination is the same. Is this a chance coincidence? It is possible that direct discrimination of numbers always has a temporal nature and that the capacity of the instantaneous memory limits the number of situations it can distinguish. In this case the limitation on spatial discrimination is explained by the hypothesis that the visual image is scanned into a time sequence (and there is a rapid switching of the eye's attention from object to object, which was discussed above) and is fed to the very same apparatus for analysis.

Be that as it may, nature has left an unfortunate gap in our mental device; therefore human beings begin work to create a "continuation of the brain" by correcting nature's mistake—humans learn to count, and thus mathematics begins.

■ COUNTING AND MEASUREMENT

FACTS TESTIFY clearly that counting emerges before the names of the numbers. In other words, the initial linguistic objects for constructing a model are not words but distinct, uniform objects: fingers, stones, knots, and lines. That is natural. During the emergence of language, words refer only to those concepts which already exist, which is to say, those which are recognized. The words "one," "two," and possibly "three" appear independently of counting (taking "counting" to mean a procedure which is prolonged in time and recognized as such) because they rely on the corresponding neuronal concepts. There is as yet nowhere from which to take the words for large numbers. To convey the size of some group of objects, the human being uses standard objects, establishing a one-to-one correspondence among them, one after the other. This is counting. When counting becomes a widespread and customary matter, word designations begin to emerge for the most frequently encountered (in other words small) groups of standard objects. Traces of their origin have remained in certain numbers. For example, the Russian word for five, *pyat,'* is suspiciously similar to the old Slavic *pyad,'* which means hand (five fingers).

There are primitive peoples who have only "one," "two," and "three"; everything else is "many." But this in no way excludes the ability to count by using standard objects, or to convey the idea of size by breaking down objects into groups of two or three, or by using as yet unreduced expressions, such as "as many as the digits on two hands, one foot, plus one." The need for counting is simply not yet great enough to establish special words. The sequence "one, two, three, many" does not reflect an inability to count to four and beyond, as is sometimes thought, but rather a distinction the human mind makes between the first three numbers and all the rest. For we can only unconsciously—and without exertion—distinguish the numbers to three. To recognize a group of four we must concentrate especially. Thus it is true for us as well as for savages that everything which is more than three is "many."

To convey large numbers people began to count in "large units": fives, tens, and twenties. In all the counting systems known to us large units are divisible by five, which indicates that the first counting tool was always the fingers. Still larger units arose from combinations of large units. Separate hieroglyphs depicting numbers up to ten million are found in Ancient Egyptian papyruses.

The beginning of measurement, just as with counting, goes back to ancient times. Measurement is already found among the primitive peoples. Measurement assumes an ability to count, and additionally it demands the introduction of a unit of measure and a measurement procedure that involves comparing what is being measured against a unit of measurement. The most ancient measures refer to the human body: pace, cubit [lit. "elbow"; the unit was the length of the forearm], and foot.

With the emergence of civilization the need for counting and for the ability to perform mathematical operations increases greatly. In developed social production, the regulation of relations among people (exchange, division of property, imposition of taxes) demands a knowledge of arithmetic and the elements of geometry. And we find this knowledge in the most ancient civilizations known to us: Babylon and Egypt.

■ NUMBER NOTATION

THE WRITING OF NUMBERS in ancient times demonstrates graphically the attitude toward the number as a direct model of reality. Let us take the Egyptian system for example. It was based on the decimal principle and contained hieroglyphs for the one (vertical line) and "large ones." To depict a number it was necessary to repeat the hieroglyph as many times as it occurred in the number. Numbers were written in a similar way by other ancient peoples (figure 9.1). The Roman system is close to this very form of notation. It differs only in that when a smaller unit stands to the left of a larger it must be subtracted. This minor refinement (together with introduction of the intermediate units V, L, and D) eliminated the necessity of writing out a series of many identical symbols, giving the Roman system such vitality that it continues to exist to the present day.

An even more radical method of avoiding the cumbersome repetition of symbols is to designate key numbers (less than 10, then even tens, hundreds, and so on) by successive letters of the alphabet. This is precisely what the Greeks did in the eighth century B.C. Their alphabet was large enough for ones, tens, and hundreds; numbers larger than 1,000 were depicted by letters with a small slash mark to the left and beneath. Thus, β signified two, $\varkappa$ signified 20, and $/\beta$ signified 2,000. Many peoples (such as the Armenians, the Jews, the Slavs) borrowed this system from the Greeks. With alphabetic nu-

	EGYPTIAN			ASSYRIAN-BABYLONIAN	PHOENICIAN	SYRIAN	PALMYRIAN	GREEK HERODIANIC	ROMAN
	HIERO-GLYPHS	HIERATIC	DEMOTIC						
1	I	I	I	▼	I	I	I	I	I
2	II	II	𝟺	▼▼	II	Ϸ	II	II	II
3	III	III	♭	▼▼▼	III	Ϸ|	III	III	III
4	IIII	ᵧ	ν:υ	▼▼▼▼	\III	ϷϷ	IIII	IIII	IV
5	III II	⅂	1	▼▼▼ ▼▼	II III	→	ᵧ	Γ	V
6	III III	⅔	ⅈ	▼▼▼ ▼▼▼	III III	↦	'ᵧ	ΓI	VI
7	IIII III	⌐	⅏	▼▼▼ ▼▼▼▼	\III III	⊢→	"ᵧ	ΓII	VII
8	IIII IIII	⫤	2	▼▼▼ ▼▼▼ ▼▼	II III III	Ϸ↦→	‴ᵧ	ΓIII	VIII
9	IIII IIIII	⅌	⅀	▼▼▼ ▼▼▼ ▼▼▼	III III III	ϷϷ↦→	‴′ᵧ	ΓIIII	IX
10	∩	∧	λ	⟨	⌐	7	Ɔ	Δ	X
11	∩I	I∧	Iλ	⟨▼	I⌐	7	'Ɔ	ΔI	XI
15	∩ III II	1∧	1λ	⟨▼▼▼	II III⌐	→	ᵞƆ	ΔΓ	XV
20	∩∩	⌒	⅃	⟨⟨	H	ο	3	ΔΔ	XX
30	∩∩∩	⅍	⅄	⟨⟨⟨	⌐H	70	Ɔ3	ΔΔΔ	XXX
40	∩∩∩∩	⇌	⌣	⟨⟨⟨⟨	HH	οο	33	ΔΔΔΔ	XL
50	∩∩∩∩∩	⇝	⅊	⟨⟨⟨⟨⟨	→ HH	700	Ɔ33	Γᴾ	L
60	∩∩∩ ∩∩∩	⅏	⅂	▼	HHH	οοο	333	ΓᴾΔ	LX
70	∩∩∩∩ ∩∩∩	⅊	⅄	▼⟨	⌐HHH	7000	Ɔ333	ΓᴾΔΔ	LXX
80	∩∩∩∩ ∩∩∩∩	⅏	⅌	▼⟨⟨	HHHH	οοοο	3333	ΓᴾΛΛΛ	LXXX
90	∩∩∩∩∩ ∩∩∩∩	⅏	⅊	▼⟨⟨⟨	⌐HHHH	70000	Ɔ33333	ΓᴾΔΔΔΔ	XC
100	9	⌐	⌐	▼►	PI	7ι	Ɔ'	H	C
200	99	⌐	⌐	▼▼▼►	PII	7ιι	Ɔ"	HH	CC
400	9999	⅏	⅏	▼▼▼▼►			Ɔ""	HHHH	CD
500	999 99	⅏	⅂³	▼▼▼▼►			Ɔᵞ	Γᴾ	D
1000	ⵁ	⅁	⅃	⟨▼►			⅁⅁'	ᵞ	M
10000	⌠			⟨⟨▼►					M
10^5	𓆰								
10^6	𓁨								
10^7	Ω								

Figure 9.1. Number notation by different ancient peoples.

meration the "model" form of the number completely disappears; it becomes merely a symbol. Simplification (for purposes of rapid writing) of characters which initially had model form leads to the same result.

"Arabic" numerals are believed to be of Indian origin, although not all specialists agree with this hypothesis. Numbers are first encountered in Indian writings in the third century B.C. At this time two forms of writing were used, Kharoshti and Brahmi, and each one had its own numerals. The Kharoshti system is interesting because the number four was selected as the intermediate stage between 1 and 10. It is likely that the oblique cross (x) used as a 4 tempted the creators of the Kharoshti numbers by its simplicity of writing while still preserving the modeling quality in full (four rays). The Brahmi numerals are more economical. It is believed that the first nine Brahmi characters finally gave rise to our modern numerals.

The loss of the model form in numbers was more than compensated for in the ancient world by the use of the abacus, a counting board with parallel grooves along which pebbles were moved. The different grooves corresponded to units of different worth. The abacus was probably invented by the Babylonians. It was used for all four arithmetic operations. Greek merchants used the abacus extensively and the same kind of counting board was common among the Romans. The Latin word for pebble, "calculus," began also to mean "computation." And the Romans conceived the idea of putting the counting pebbles on rods, which is how the abacus still in use today originated. These very simple counting devices were enormously important and only gave way to computations on slateboards or paper after the positional system of notation had completely formed.

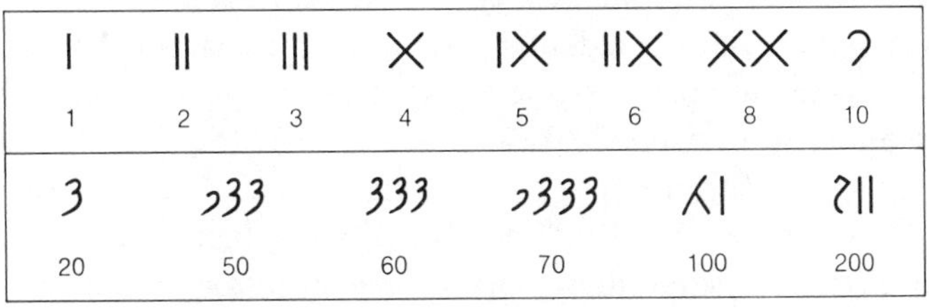

I	II	III	X	IX	IIX	XX	?
1	2	3	4	5	6	8	10

3	?33	333	?333	ΛI	?II
20	50	60	70	100	200

Figure 9.2. Kharoshti numerals.

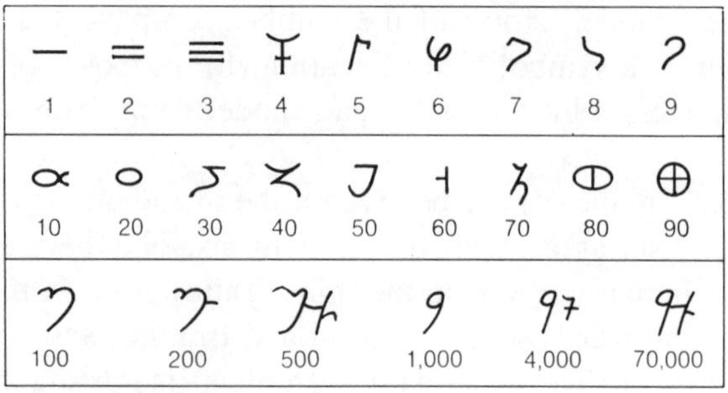

Figure 9.3. Brahmi numerals.

■ THE PLACE-VALUE SYSTEM

THE BABYLONIANS laid the foundations of the place-value system. In the number system they borrowed from the Sumerians, we see two basic "large ones": ten and sixty, from the most ancient clay tablets which have come down to us, dating to the beginning of the third millennium B.C. We can only guess where the number sixty was taken from. The well-known historian of mathematics O. Neigebauer believes that it originated in the relation between the basic monetary units in circulation in Mesopotamia: one *mana* (in Greek *mina*) was sixty *shekels*. Such an explanation does not satisfy our curiosity because the question immediately arises: why are there sixty shekels in a mana? Isn't it precisely because a system based on sixty was in use? After all, we don't count by tens and hundreds because there are 100 kopecks in a ruble! F. Thureau-Dangin, an Assyriologist, gives linguistic arguments to show that the number system was the primary phenomenon and the system of measures came second. Selection of the number sixty was apparently a historical accident, but one can hardly doubt that this accident was promoted by an important characteristic of the number sixty, namely that it has an extraordinarily large number of divisors 2, 3, 4, 5, 6, 10, 12, 15, 20, and 30). This is a very useful feature both for a monetary unit (since its existence, money has been evenly subdivided) and for establishing a system of counting (if we assume that some wise man introduced it, guided by considerations of convenience in calculation).

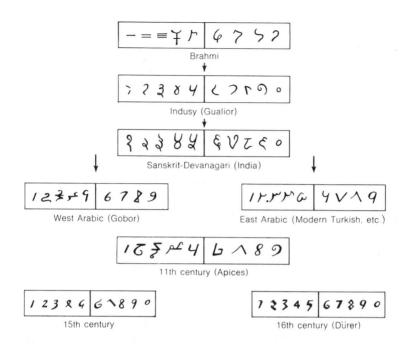

Figure 9.4. The genealogy of modern numerals (according to Menninger, Zohlwort, and Ziffer).

The mathematical culture of the Babylonians is known to us from texts dating from the Ancient Bablonian (1800–1600 B.C.) and the Seleucidae epoch (305–64 B.C.). A comparison of these texts shows that no radical changes took place in the mathematics of the Babylonians during this time.

The Babylonians depicted 1 by a narrow vertical wedge ▼, while 10 was a wide horizontal wedge ◀ . The number 35 looked like this ◀◀◀ ▼▼▼ . All numbers up to 59 were represented analogously. But 60 was depicted once again by a narrow vertical wedge, just the same as 1. In the most ancient tablets it can be seen that the wedge representing 60 is larger than the wedge for 1. Thus, the number 60 was not only understood as a "large one" but was so represented. "Large tens" appeared correspondingly for large units multiplied by 10. Later, the difference between large and small wedges was lost and they began to be distinguished by their position. In this way the positional system arose. A Babylonian would write the number $747 = 12 \times 60 + 27$ in the form ⟨▼▼⟨⟨▼▼▼ . The third 60-base position corresponds to the number $60^2 = 3{,}600$ and so on. But the most remarkable thing is that the Babylonians also represented fractions in this way. In a number following the number of ones, each unit signified $1/60$, in the next number each unit was $1/3{,}600$, and so on. In modern decimal notation we separate the whole part from the fraction part by a period or comma. The Babylonians did not. The number ▼⟨⟨⟨ could signify 1.5 or 90 with equal success. This same uncertainty occurred in writing whole numbers: the numbers n, $n \times 60$, $n \times 60^2$, and so on were indistinguishable. Multipliers and divisors divisible by 60 had to be added according to the sense. Because 60 is a fairly large number, this did not cause any particular problems.

When we compare the Babylonian positional system with our modern one we see that the uncertainty in the multiplier 60 is a result of the absence of a character for zero, which we would add the necessary number of times at the end of a whole number or the beginning of a fraction. Another result of the absence of the zero is an even more serious uncertainty in interpreting a numerical notation that in our system requires a zero in an intermediate position. In the Babylonian notation, how can the number $3{,}601 = 1 \cdot 60^2 + 0 \cdot 60 + 1$ be distinguished from the number $61 = 1 \cdot 60 + 1$? Both of these numbers are represented by two units (ones). Sometimes this kind of uncertainty was eliminated by separating the numbers, leaving an empty place for the missing position. But this method was not used systematically, and in many cases a large gap between numbers did not mean anything. In the astronomical tables of the Seleucidae epoch one finds the missing position designated by means of a character resembling our period. We do not find anything of the sort in the Ancient Babylonian epoch. But how were the ancient Babylonians able to avoid confusion?

The solution to the riddle is believed to consist in the following.[1] The early Babylonian mathematical texts which have come down to us are collections of problems and their solutions, unquestionably created as learning aids. Their purpose was to teach practical methods of solving problems. But not one of the texts describes how to perform arithmetic operations, in particular the operations, complex for that time, of multiplication and division. Therefore, we assume that the students knew how to do them. It is improbable that they performed the computations in their head; they probably used some abacus-like calculating device. On the abacus, numbers appear in their natural, spontaneously positioned form and no special character for the zero is needed; the groove corresponding to an empty position simply remains without pebbles. Representation of a number on the abacus was the basic form of assigning a number, and there was no uncertainty in this representation. The numbers given in cuneiform mathematical texts serve as answers to stage-by-stage calculations, so that they could be used to check correctness during the solution. The student made the calculations on the abacus and checked them against the clay tablet. Clearly the absence of a character for empty positions did not hinder such checking at all. When voluminous astronomical tables became widespread and were no longer used for checks but rather as the sole source of data, a separation sign began to be used to represent the empty positions. But the Babylonians never put their own ''zero'' at the end of a word; it is obvious that they perceived it only as a separator, and not as a number.

Having familiarized themselves with both the Egyptian and the Babylonian systems of writing fractions and performing operations on them, the Greeks selected the Babylonian system for astronomical calculations because it was incomparably better, but they preserved their own alphabetic system for writing whole numbers. Thus the Greek system used in astronomy was a mixed one: the whole part of the number was represented in the decimal nonpositional system while the fractional part was in the 60-base positional system—not a

[1] See the remarks by I. N. Veselovsky on the Russian translation of B. L. van der Waerden's book *Ontwakende Wetenschap* (Russian *Probuzhdayushchayasya nauka,* Fizmatgiz Publishing House, 1959; English *Science Awakening,* New York: Oxford University Press, 1969).

very logical solution by the creators of logic! Following their happy example we continue today to count hours and degrees (angular) in tens and hundreds, but we divide them into minutes and seconds.

The Greeks did introduce the modern character for zero into the positional system, deriving it—in the opinion of a majority of specialists—from the first letter of the word ουδεν, which means "nothing." In writing whole numbers (except for the number 0) this character was not used, naturally, because the alphabetic system which the Greeks used was not a positional one.

The modern number system was invented by the Indians at the beginning of the sixth century A.D. They applied the Babylonian positional principle and the Greek character for zero to designate an empty place to a base of 10, not 60. The system proved to be consistent, economical, not in contradiction with tradition, and extremely convenient for computations.

The Indians passed their system on to the Arabs. The positional number system appeared in Europe in the twelfth century with translations of al-Kwarizmi's famous Arab arithmetic. It came into bitter conflict with the traditional Roman system and in the end won out. As late as the sixteenth century, however, an arithmetic textbook was published in Germany and went through many editions using exclusively "German," which is to say Roman, numerals. It would be better to say "numbers," because at that time the word "numerals" was used only for the characters of the Indian system. In the preface of this textbook the author writes: "I have presented this arithmetic in conventional German numbers for the benefit and use of the uneducated reader (who will find it difficult to learn numerals at the same time)." Decimal fractions began to be used in Europe with Simon Stevin (1548–1620).

■ APPLIED ARITHMETIC

THE MAIN LINE to modern science lies through the culture of Ancient Greece, which inherited the achievements of the ancient Egyptians and Babylonians. The other influences and relations (in particular the transmission function carried out by the Arabs) were of greater or lesser importance but, evidently, not crucial. The sources of the Egyptian and Sumerian-Babylonian civilizations are lost in the dark of primitive cultures. In our review of the history of science, therefore, we shall limit ourselves to the Egyptians, Babylonians, and Greeks.

We have already discussed the number notation of the Egyptians and Babylonians. All we need now is to add a few words about how the Egyptians wrote fractions. From a modern point of view their system was very original, and very inconvenient. The Egyptians had a special form of notation used only for so-called "basic" fractions, that is, those obtained by dividing one by a whole number; in addition they used two simple fractions which had had special hieroglyphs from ancient times: $^2/_3$ and $^3/_4$. In the very latest papyruses, however, the special designation for $^3/_4$ disappeared. To write a basic fraction the symbol ➴, which meant "part," had to be placed above a conventional number (the denominator). Thus ⬛ⅲ $= ^1/_{12}$.

The Egyptians expanded the other fractions into the sum of several basic fractions. For example, $^3/_8$ was written as $^1/_4 + ^1/_8$, and $^2/_7$ was written as $^1/_4 + ^1/_{28}$. For the result of dividing 2 by 29 an Egyptian table gave the following expansion:

$$\frac{2}{29} = \frac{1}{29} + \frac{1}{58} + \frac{1}{174} + \frac{1}{232}.$$

We are not going to dwell on the computational techniques of the Egyptians and Babylonians. It is enough to say that they both were able to perform the four arithmetic operations on all numbers (whole, fractional, or mixed) which they met in practice. They used auxiliary mathematical tables for operations with fractions; these were tables of inverse numbers among the Babylonians and tables of the doubling of basic fractions among the Egyptians. The Egyptians wrote intermediate results on papyrus, whereas the Babylonians apparently performed their operations on an abacus and thus the details of their technique remain unknown.

What did the ancient mathematicians calculate? One fragment of an Egyptian papyrus from the times of the New Empire (1500–500 B.C.) describes the activity of the pharaoh's scribes very colorfully and with a large dose of humor; for this reason it is invariably cited in all books on the history of mathematics, and we shall not be an exception. Here is the excerpt, somewhat shortened:

> I will cause you to know how matters stand with you, when you say "I am the scribe who issues commands to the army." . . . I will cause you to be abashed when I disclose to you a command of your lord, you, who are his Royal Scribe. . . . the clever scribe who is at the head of the troops. A building ramp is to be constructed, 730 cubits long, 55 cubits wide, containing 120 compartments, and filled with reeds and beams; 60 cubits high at its summit, 30 cubits in the middle, with a batter of twice 15 cubits and its pavement 5 cubits. The quantity of bricks needed for it is asked of the generals, and the scribes are all asked together, without one of them knowing anything. They all put their trust in you and say "You are the clever scribe, my friend! Decide for us quickly!" Behold your name is famous. . . . Answer us how many bricks are needed for it? [2]

Despite its popularity, this text is not too intelligible. But it nevertheless does give an idea of the problems Egyptian scribes had to solve.

Specifically, we see that they were supposed to able to calculate areas and volumes (how accurately is another question). And in fact, the Eyptians possessed a certain knowledge of geometry. According to the very sound opinion of the Ancient Greeks, this knowledge arose in Egypt itself. One of the philosophers of Aristotle's school begins his treatise with the words:

> Because we must survey the beginning of the sciences and arts here we will state that, according to the testimony of many, geometry was discovered by the Egyptians and originated during the measurement of land. This measurement was necessary because the flooding of the Nile River constantly washed away the boundaries. There is nothing surprising in the fact that this science, like others, arose from human need. Every emerging knowledge passes from incomplete to complete. Originating through sensory perception it increasingly becomes an object of our consideration and is finally mastered by our reason.[3]

---

[2] van der Waerden, *Science Awakening,* p 17.
[3] This fragment has come down to us through Procul (fifth century B.C.), a commentator on Euclid.

The division of knowledge into incomplete and complete and a certain apologetic tone concerning the "low" origin of the science are, of course, from the Greek philosopher. Neither the Babylonians nor the Egyptians had such ideas. For them knowledge was something completely homogeneous. They were able to make geometric constructions and knew the formulas for the area of a triangle and circle just as they knew how to shoot bows and knew the properties of medicinal plants and the dates of the Nile's floods. They did not know geometry as the art of deriving "true" formulas; among them it existed, as B. Van der Warden expressed it, only as a division of *applied arithmetic*. It is obvious that they employed certain guiding considerations in obtaining the formulas, but these considerations were of little interest to them. They did not affect their attitude toward the formula.

■ THE ANCIENTS' KNOWLEDGE OF GEOMETRY

WHAT GEOMETRY did the Egyptians know?—the correct formulas for the area of a triangle, a rectangle, and a trapezium. The area of an irregular quadrangle, to judge by the one remaining document, was calculated as follows: half the sum of two opposite sides was multiplied by half the sum of the other two opposite sides. This formula is grossly wrong (except where the quadrangle is rectangular, in which case the formula is unnecessary). There is no reasonable sense in which it can even be called approximate. It appears that this is the first historically recorded example of a proposition which is derived from "general considerations," not from a comparison with the data of experience. The Egyptians calculated the area of a circle by squaring $8/9$ of its diameter, a difference of about 1 percent from the value of π.

They calculated the volumes of parallelepipeds and cylinders by multiplying the area of the base by the height. The most sophisticated achievement of Egyptian geometry known to us is correct computation of the volume of a truncated pyramid with a square base (the Moscow papyrus). It follows the formula

$$V = (a^2 + ab + b^2)\frac{h}{3}$$

where h is the height, a and b are the sides of the upper and lower bases.

We have only fragmentary information on the Ancient Babylonians' knowledge of mathematics, but we can still form a general idea of it. It is completely certain that the Babylonians were aware of what came to be called the "Pythagorian theorem"—the sum of the squares of the sides of a right triangle is equal to the square of the hypotenuse. Like the Egyptians they computed the areas of triangles and trapeziums correctly. They computed the circumference and area of a circle using a value of $\pi = 3$, which is much worse than the Egyptian approximation. The Babylonians calculated the volume of a truncated pyramid or cone by multiplying half of the sum of the areas of the bases by the height (an incorrect formula).

■ A BIRD'S EYE VIEW OF ARITHMETIC

THE SITUATIONS and representations in the human nervous system model the succession of states of the environment. Linguistic objects model the succession of situations and representations. As a result a theory is a "two-story" linguistic model of reality (see figure 9.5). This diagram shows the use of a theory. The situation S_1 is encoded by linguistic object L_1. This object may, of course, consist of a set of other objects and may have a very highly complex structure. Object

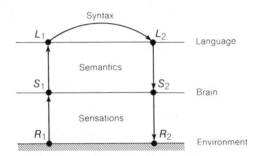

Figure 9.5. Diagram of the use of a linguistic model of reality.

L_1 is the *name* for S_1. Sometime later situation S_1 changes into situation S_2. A certain linguistic activity is performed and we convert L_1 into another object L_2; if our model is correct L_2 is the name of S_2. Then, without knowing the real situation S_2, we can get an idea of it by *decoding* linguistic object L_2. The linguistic model is plainly determined by both the semantics of objects L_1 (the "material part" according to Russian military terminology) and the type of linguistic activity which converts L_1 into L_2.

Notice that we have not said anything about "isolating the essential aspects of the phenomenon," "the cause-effect relation," or other such things which are usually set in places of honor when describing the essence of scientific modeling. And in our presentation, situation S_1 does not "generate" situation S_2 but only "changes into" it. Of course, this is no accident. The diagram we have drawn logically precedes the above-mentioned philosophical concepts. If we have a linguistic model (and only to the extent that we do have one) we can talk about the essential aspect of a phenomenon, idealization, the cause-effect relation, and the like. Although they appear to be conditions for the creation of a linguistic model, all of these concepts are in fact nothing but description in general terms (although very important and necessary ones) of already existing models. Although these concepts appear to "explain" why a linguistic model can exist in general, in reality they are elements of a linguistic model of the next level of the control hierarchy and, of course, appear later in history than the primary linguistic models (for example arithmetic ones). Before using these concepts, therefore, we must ascertain that linguistic models exist in general. And on this level of description we need not add anything to the diagram shown in figure 9.5.

But theories are created and developed by the trial and error method. If there is a starting point, then beginning from it a person tries to build linguistic constructions and test the results. The phases of building and testing are constantly alternating: construction gives rise to testing and testing gives rise to new constructions.

The starting point of arithmetic is the concept of the whole number. The aspect of reality this concept reflects is the following: the relation of the whole to its parts, the procedure for breaking the whole down into parts. The same thought can be expressed the other way around: a number is a procedure for joining parts into a whole, that is, into a certain set. Two numbers are considered identical if their parts (set elements) can be placed in a one-to-one correspondence; establishing this correspondence is counting. It is obvious, however, that numbers are not enough for a theory; we must also have operations with them. These are the elements of the model's functioning, the conversions $L_1 \rightarrow L_2$. Let us take two numbers n and m and represent them schematically as two modes of breaking a whole down into parts (figure 9.6a).

How can we from these two numbers obtain a third—that is, a third mode of breaking down the whole into parts? Two modes come to mind immediately. They can be called parallel and sequential joining of breakdowns. In the parallel mode both wholes form parts of a new whole (figure 9.6b). This breakdown (number) we call the *sum* of the two numbers. With the sequential mode we take one of the breakdowns and break down each of its parts in accordance with the other breakdown (figure 9.6c). The new number is called the *product*. It does not depend on the order of the generating numbers. This can be seen very well if we interpret the actions with the numbers not as joining breakdowns but as forming a new set. The sum is obviously the result of merging the two sets into one (their union). The prototype of the product is the set of combinations of any element of the first set with any element of the second (in mathematics such a set is called the *direct product* of sets). The connection between this definition and the preceding one can be traced as follows. Suppose the first breakdown divides whole number A into parts $a_1, a_2, \ldots, a_n$ and the second divides B into parts $b_1, b_2, \ldots, b_m$. After performing the first breakdown we mark the parts obtained with the letters a_i. Breaking down each part into parts b_j we keep the first letter and add a second. This means that in each part of the result there will be an $a_i b_j$, and all these combinations will be different. The approaches from the whole to the part and from the part to the whole complement one another. It is also easy to see from figure 9.6c that multiplication can be reduced to repeated addition.

Of course the ancients who were creating arithmetic were far from this reasoning. But then again, the frog did not know that its nervous system had to be organized on the hierarchical principle either! What is important is that *we* know this.

Having linguistic objects that depict numbers and being able to perform addition and multiplication with them we receive a theory that gives us working models of reality. Let us figure a very simple example, which clarifies the diagram in figure 9.6.

Suppose a certain farmer has planted wheat in a field 60 paces long and 25 paces wide. We shall assume that the farmer expects a yield of one bushel of wheat per square pace. Before beginning the harvest he wants to know how many bushels of wheat he will get. In this case S_1 is the situation before the wheat harvest, specifically including the result of measuring the width and length of the field in paces and the expected yield; S_2 is the situation after the harvest, specifically including the result of measuring the amount of wheat in bushels; L_1 is the linguistic object 60×25 (the multiplication sign is a reflection of situation S_1 just as the numbers 60 and 25 are; it reflects the structure of the set of square paces on the plane as a direct product of the sets of linear paces for length and width); L_2 is the linguistic object 1,500.

Note that by *theory* we mean simply a linguistic model of reality which gives something new in comparison with neuronal models. This definition does not take into account that theories may form a control hierarchy; this fact is difficult to reflect without introducing mathematical apparatus. More general models can generate more particular ones. We shall consider the terms *theory* and *linguistic model*

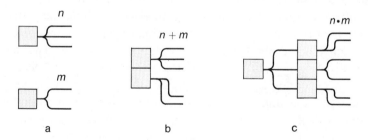

Figure 9.6. Operations on whole numbers.

to be synonymous, but nonetheless when we are speaking of one model generating another, we shall call the more general one a theory and the more particular one a model.

■ REVERSE MOVEMENT IN A MODEL

A THEORY THAT has just been created must first be tested comprehensively. It must be compared with experience and searched for flaws. If the theory is valid, an attempt must then be made to give the model "reverse movement," to determine specific characteristics of L_1 on the basis of a given L_2. This procedure is by no means without practical importance. A person uses a model for planning purposeful activity and wants to know what he must do to obtain the required result—what L_1 should be in order to obtain a given L_2. In our example with the farmer, the question can be put as follows: given the width of a certain field, what should the length be to obtain a given amount of wheat?

But studying the reverse movement of a model is not always dictated by practical needs of the moment. Often it is done for pure curiosity—to "see what happens." Nonetheless, the result of such activity will be a better understanding of the organization and characteristics of the model and the creation of new constructions and models which, in the last analysis, will lead to greatly enlarged practical usefulness. This is the supreme wisdom of nature, which gave human beings "pure" curiosity.

In arithmetic the reverse movement of a model leads to the concept of the *equation*. The simplest equations generate the operations of subtraction and division. Using modern algebraic language, we define the difference $b - a$ as the solution of the equation $a + x = b$—in which x is the number that makes this equality true. The quotient from dividing b by a is determined analogously. The operation of division generates a new construction: the fraction. Repeated multiplication of a number by itself generates the construction of exponential degree, and the reverse movement generates the operation of extracting the root. This completes the list of arithmetic constructions in use among the Ancient Egyptians and Babylonians.

■ SOLVING EQUATIONS

WITH THE DEVELOPMENT of techniques of computation, and with the development of civilization in general, increasingly complex equations began to appear and to be solved. The ancients, of course, did not know modern algebraic language. They expressed equations in ordinary conversational language as is done in our grammar school arithmetic textbooks. Nevertheless, their substance was not changed. The ancients (and today's school children) were solving equations.

The Egyptians called the quantity subject to determination the *akha,* which is translated as "certain quantity" or "bulk." Here is an example of the wording of a problem from an Egyptian papyrus: "A quantity and its fourth part together give 15." In modern mathematical terminology this is the problem of "parts," and in algebraic language it corresponds to the equation $x + {}^1/_4 x = 15$.

Let us give an example of a more complex problem from Egyptian times.

"A square and another square whose side is ${}^1/_2 + {}^1/_4$ of a side of the first square together have an area of 100. Calculate this for me."

The solution in modern notation is as follows:

$$x^2 + \left(\frac{3}{4}x\right)^2 = 100$$

$$\left(1 + \frac{9}{16}\right)x^2 = 100$$

$$\frac{5}{4}x = 10$$

$$x = 8, \quad \frac{3}{4}x = 6$$

Here is the description of the solution in a papyrus:

"Take a square with side 1 and take ${}^1/_2 + {}^1/_4$ of 1, that is, ${}^1/_2 + {}^1/_4$, as the side of the second area. Multiply ${}^1/_2 + {}^1/_4$ by itself; this gives ${}^1/_2 + {}^1/_{16}$. Because the side of the first area is taken as 1 and the second as ${}^1/_2 + {}^1/_4$, add both areas together; this gives $1 + {}^1/_2 + {}^1/_{16}$. Take the root from this: this will be $1 + {}^1/_4$. Take the root of the given 100: this will be 10. How many times does $1 + {}^1/_4$ go into 10? It goes eight times."

The rest of the text has not been preserved, but the conclusion is obvious: $8 \cdot 1 = 8$ is the side of the first square and $8 \cdot (^1/_2 + {}^1/_4) = 6$ is the second.

The Egyptians were able to solve only linear and very simple quadratic equations with one unknown. The Babylonians went much further. Here is an example of a problem from the Babylonian texts: "I added the areas of my two squares: $25 \cdot \frac{25}{60}$. The side of the second square is $^2/_3$ of the side of the first and five more." This is followed by a completely correct solution of the problem. This problem is equivalent to a system of equations with two unknowns:

$$x^2 + y^2 = 25 \cdot \frac{25}{60}$$
$$y = \frac{2}{3}x + 5$$

The Babylonians were able to solve a full quadratic equation:

$$x^2 \pm ax = b,$$

cubic equations:

$$x^3 = a$$
$$x^2(x + 1) = a$$

and systems of equations similar to those given above as well as ones of the type

$$x \pm y = a, \; Xy = b$$

In addition to this they used formulas

$$(a + b)^2 = a^2 + 2ab + b^2$$
$$(a + b)(a - b) = a^2 - b^2$$

were able to sum arithmetic progressions, knew the sums of certain number series, and knew the numbers which later came to be called Pythagorian (such whole numbers x, y, and z that $x^2 + y^2 = z^2$).

■ THE FORMULA

THE PLACE of Ancient Egypt and Babylon in the history of mathematics can be defined as follows: the *formula* first appeared in these cultures. By formula we mean not only the alphanumeric expression of modern algebraic language but in general any linguistic object which is an exact (formal) prescription for how to make the conversion $L_1 \rightarrow L_2$ or any auxiliary conversions within the framework of language. Formulas are a most important part of any elaborated theory although, of course, they do not exhaust it because a theory also includes the meanings of linguistic objects L_i. The assertion that there is a relation between the magnitudes of the sides in a right triangle, which is contained in the Pythagorian theorem, is a formula even though it is expressed by words rather than letters. A typical problem with a description of the process of solution ("Do it this way!") and with a note that the numbers may be arbitrary (this may not be expressed but rather assumed) is also a formula. It is precisely such formulas which have come down to us in the Egyptian papyruses and the Babylonian clay tablets.

CHAPTER TEN
From Thales to Euclid

■ PROOF

NEITHER IN Egyptian nor in Babylonian texts do we find anything even remotely resembling mathematical proof. This concept was introduced by the Greeks, and is their greatest contribution. It is obvious that some kind of guiding considerations were employed earlier in obtaining new formulas. We have even cited an example of a grossly incorrect formula (for the area of irregular quadrangles among the Egyptians) which was plainly obtained from externally plausible "general considerations." But only the Greeks began to give these guiding considerations the serious attention they deserved. The Greeks began to analyze them from the point of view of how convincing they were, and they introduced the principle according to which every proposition concerning mathematical formulas, with the exception of just a small number of "completely obvious" basic truths, must be proved—derived from these "perfectly obvious" truths in a convincing manner admitting of no doubt. It is not surprising that the Greeks, with their democratic social order, created the doctrine of mathematical proof. Disputes and proofs played an important part in the life of the citizens of the Greek city-state (*polis*). The concept of proof already existed; it was a socially significant reality. All that remained was to transfer it to the field of mathematics, which was done as soon as the Greeks became acquainted with the achievements of the ancient Eastern civilizations. It must be assumed that a certain part here was also played by the role of the Greeks as young, curious students in relation to the Egyptians and Babylonians, their old teachers who did not always agree with one another. In fact, the Babylonians determined the area of a circle according to the formula $3r^2$, while the Egyptians used the formula $(\frac{8}{9} \cdot 2r)^2$. Where was the truth? This was something to think about and debate.

The creators of Egyptian and Babylonian mathematics have remained anonymous. The Greeks preserved the names of their wise men. The first, Thales of Miletus, is also the first name included in the history of science. Thales lived in the sixth century B.C. in the city of Miletus on the Asia Minor coast of the Aegean Sea. One date in his life has been firmly established: in 585 B.C. he predicted a solar eclipse—unquestionable evidence of Thales's familiarity with the culture of the ancient civilizations, because the experience of tens and hundreds of years is required to establish the periodicity of eclipses. Thales had no Greek predecessors, and could therefore only have taken his knowledge of astronomy from the scientists of the East.

Thales, the Greeks assert, gave the world the first mathematical proofs. Among the propositions (theorems) proved by him they mention the following:

1 The diameter divides a circle into two equal parts.
2 The base angles of an isosceles triangle are equal.
3 Two triangles which have an identical side and identical angles adjacent to it are equal.

In addition, Thales was the first to construct a circle circumscribed about a right triangle (and it is said that he sacrificed an ox in honor of this discovery).

The very simple nature of these three theorems and their intuitive obviousness shows that Thales was entirely aware of the importance of proof as such. Plainly, these theorems were proved not because there was doubt about their truth but in order to make a beginning at systematically finding proof and developing a technique for proof. With such a purpose it is natural to begin by proving the simplest propositions.

Suppose triangle *ABC* is isosceles (see figure 10.1), which is to say side *AB* is equal to side *BC*. Let us divide angle *ABC* into two equal parts by line *BD*. Let us mentally fold our drawing along line *BD*. Because angle *ABD* is equal to angle *CBD*, line *BA* will lie on line *BC;* and because the length of the segments *AB* and *BC* is equal, point *A* will lie on point *C*. Because point *D* remains in place, angles *BCD* and *BAD* must be equal. Whereas formerly it only *seemed* to us that angles *BCD* and *BAD* were equal (Thales probably spoke this way to his fellow citizens), we have now *proved* that these angles necessarily and with absolute precision must be equal (the Greeks said "similar") to one another; that is, they match when one is placed on the other.

The problem of construction is more complex and here the result is not at all obvious beforehand. Let us draw a right triangle (see figure 10.2). May a circle be drawn such that all three vertices of the triangle appear on it? And if so, how? It is not clear. But suppose that intuition suggests a solution to us. We divide the hypotenuse *BC* into two equal segments at point *D*. We connect it with point *A*. If segment *AD* is equal in magnitude to segment *DC* (and therefore also to *BD*) we can easily draw the required circle by putting the point of a compass at point *D* and taking segment *DC* as the radius. But is it true that *AD* = *DC*, that is to say triangle *ADC* is an isosceles triangle? It is not clear. It seems probable, but in any case it is far from obvious. Now we shall take the crucial step. We shall add point *E* to our triangle, making rectangle *ABEC*, and draw in a second diagonal

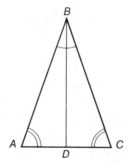

Figure 10.1. Isosceles triangle.

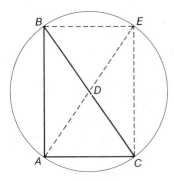

Figure 10.2. Construction of a circle
described around a right triangle.

AE. Suddenly it becomes obvious that triangle ADC is isosceles. Indeed, from the overall symmetry of the drawing it is clear that the diagonals are equal and intersect at the point which divides them in half—at point D. We have not yet arrived at proof, but we already are at that level of clarity where formal completion of the proof presents no difficulty. For example, relying on the equality of the opposite sides of the rectangle (which can be derived from even more obvious propositions if we wish), we complete the proof by the following reasoning: triangles ABC and AEC are equal because they have side AC in common, sides AB and EC are equal, and angles BAC and ECA are right angles; therefore angle EAC is equal to angle BCA. That is, triangle ADC is an isosceles triangle, which is what had to be proved.

■ THE CLASSICAL PERIOD

SO, FROM a few additional points and lines on a drawing, a chain of logical reasoning, and simple and obvious truths we receive truths which are by no means simple and by no means obvious, but whose correctness no one can doubt for a minute. This is worth sacrificing an ox to the gods for! One can imagine the delight the Greeks experienced upon making such a discovery. They had struck a vein of gold and they diligently began working it. In the time of Pythagoras (550

B.C.) the study of mathematics was already very widespread among people who had leisure time and was considered a noble, honorable, and even sacred matter. Advances and discoveries, each more marvelous than the one before, poured from the horn of plenty.

The appearance of proof was a metasystem transition within language. The formula was no longer the apex of linguistic activity. A new class of linguistic objects appeared, proof, and there was a new type of linguistic activity directed to the study and production of formulas. This was a new stage in the control hierarchy and its appearance called forth enormous growth in the number of formulas (the law of branching of the penultimate level).

The metasystem transition always means a qualitative leap forward—a flight to a new step, swift, explosive development. The mathematics of the countries of the Ancient East remained almost unchanged for up to two millennia, and a person of our day reads about it with the condescension of an adult toward a child. But in just one or two centuries the Greeks created all of the geometry our high school students sweat over today. Even more, for the present-day geometry curriculum covers only a part of the achievements of the initial, "classical," period of development of Greek mathematics and culture (to 330 B.C.). Here is a short chronicle of the mathematics of the classical period.

585 B.C. *Thales of Miletus*. The first geometric theorems.

550 B.C. *Pythagoras* and his followers. Theory of numbers. Doctrine of harmony. Construction of regular polyhedrons. Pythagorean theorem. Discovery of incommensurable line segments. Geometric algebra. Geometric construction equivalent to solving quadratic equations.

500 B.C. *Hippasas*, Pythagorean who was forced to break with his comrades because he shared his knowledge and discoveries with outsiders (this was forbidden among the Pythagoreans). Specifically, he gave away the construction of a sphere circumscribed about a dodecahedron.

430 B.C. *Hippocrates of Chios* (not to be confused with the famous doctor Hippocrates of Kos). He was considered the most famous geometer of his day. He studied squaring the circle, making

complex geometric constructions. He knew the relationship between inscribed angles and arcs, the construction of a regular hexagon, and a generalization of the Pythagorean theorem for obtuse- and acute-angled triangles. Evidently, he considered all these things elementary truths. He could square any polygon, that is, construct a square of equal area for it.

427–348 B.C. *Plato*. Although Plato himself did not obtain new mathematical results, he knew mathematics and it sometimes played an important part in his philosophy—just as his philosophy played an important part in mathematics. The major mathematicians of his time, such as Archytas, Theaetetus, Eudoxus, were Plato's friends; they were his students in the field of philosophy and his teachers in the field of mathematics.

390 B.C. *Archytas of Tarentum*. Stereometric solution to the problem of doubling the cube—that is, constructing a cube with a volume equal to twice the volume of a given cube.

370 B.C. *Eudoxus of Cnidus*. Elegant, logically irreproachable theory of proportions closely approaching the modern theory of the real number. The "exhaustion method," which forms the basis of the modern concept of the integral.

384–322 B.C. *Aristotle*. He marked the beginning of logic and physics. Aristotle's works reveal a complete mastery of the mathematical method and a knowledge of mathematics, although he, like his teacher Plato, made no mathematical discoveries. Aristotle the philosopher is inconceivable without Aristotle the mathematician.

300 B.C. *Euclid*. Euclid lived in a new and different age, the Alexandrian Epoch. In his famous *Elements* Euclid collected and systematized all the most important works on mathematics which existed at the end of the fourth century B.C. and presented them in the spirit of the Platonic school. For more than 2,000 years school courses in geometry have followed Euclid's *Elements* to some extent.

■ PLATO'S PHILOSOPHY

WHAT IS MATHEMATICS? What does this science deal with? These questions were raised by the Greeks after they had begun to construct the edifice of mathematics on the basis of proofs, for the aura of absolute validity, of virtual sanctity, which mathematical knowledge acquired thanks to the existence of the proofs immediately made it stand out against the background of other everyday knowledge. The answer was given by the Platonic *theory of ideas*. This theory formed the basis of all Greek philosophy, defined the style and way of thinking of educated Greeks, and exerted an enormous influence on the subsequent development of philosophy and science in the Greco-Roman-European culture.

It is not difficult to establish the logic which led Plato to his theory. What does mathematics talk about? About points, lines, right triangles, and the like. But are there in nature points which do not have dimensions? Or absolutely straight and infinitely fine lines? Or exactly equal line segments, angles, or areas? It is plain that there are not. So mathematics studies nonexistent, imagined things; it is a science about nothing. But this is completely unacceptable. In the first place, mathematics has unquestionably produced practical benefits. Of course, Plato and his followers despised practical affairs, but this was a logical result of philosophy, not a premise. In the second place, any person who studies mathematics senses very clearly that he is dealing with reality, not with fiction, and this sensation cannot be rooted out by any logical arguments. Therefore, the objects of mathematics really exist but not as material objects, rather as images or ideas, because in Greek the word "idea" ($\iota\delta\epsilon\alpha$) in fact meant "image" or "form." [1] Ideas exist outside the world of material things and independent of it. Material things perceived by the senses are only incomplete and temporary copies (or shadows) of perfect and eternal ideas. The assertion of the real, objective existence of a world of ideas is the essence of Plato's teaching ("Platonism").

For many centuries hopelessly irresolvable disputes arose among the Platonists over attempts to in some way give concrete form to the notion of the world of ideas and its interaction with the material world. Plato himself wisely remained invulnerable, avoiding specific,

---

[1] The resemblance in sound between the Greek $\iota\delta\epsilon\alpha$ and the Russian *vid* is not accidental; they come from a common Indo-European root. (Compare also Latin "vidi"—past tense of "to see.")

concrete terms and using a metaphorical and poetic language. But he did have to enter a polemic with his student Eudoxus, who not only proved mathematical theorems but also defended the assertion that ideas are "admixed" with things perceived by the senses, determining their attributes.

The concepts of mathematics are not the only inhabitants of Plato's "world of ideas"; every general concept claims a place in it. The reasoning which substantiates this claim is as follows. Our language has words and phrases to signify unique concepts such as proper names: the island of Samos, Athens, Hippocrates. We get these concepts from sense perception of the corresponding things. But we also have general concepts such as human being, tree, and the like. Where do we get these? After all, sense perception gives us only concrete concepts: a particular person, a particular tree, and so on. If things generate concrete concepts in us, what generates general concepts? Plato's answer was "ideas": the idea of a human being, the idea of a tree, and so on.

The existence of the world of ideas secures for mathematics a strong and lofty position; it becomes the science of ideas. Sensory experience gives us imperfect, approximate knowledge of imperfect, approximate embodiments of ideas. Mathematical proofs give perfect knowledge of the ideas themselves. "By means of mathematics," Plato writes, "the organ of the soul is purged and receives new vital force, while other occupations destroy the soul and deprive it of its ability to see; yet it is much more valuable than a thousand eyes because only through it can the truth be revealed."

Under the influence of Plato's idealism the mathematicians of Ancient Greece tried to banish from their science everything that could be interpreted as a reference to the data of sensory experience. This had positive consequences, because it promoted the development of techniques for proof and led to the creation of the concept of deductive theory. The Greeks tried to make proofs logically irreproachable, to exclude from them doubtful arguments and implicit assumptions which appealed to perception. They did not demonstrate, they proved. They strove to reduce the number of explicit assumptions to a minimum and leave only those which could be considered an expression of the attributes of the "ideas themselves," but not of the things—which is to say the attributes that reveal themselves to reason, to the "internal view," but not to the sense organs. These assumptions were included in the definitions of the initial concepts or, to be more precise, *words,* because to the Greek concepts (ideas) existed as objective reality independent of any words; definitions were only needed to avoid mistakes in establishing correspondences between words and concepts. This was because the explicit assumptions made by the Greek mathematicians seemed to them not definitions in the modern sense of the word (according to which the definition generates the mathematical object) but simply designations of those *true attributes of actually existing ideas* which can be comprehended by reason more easily than others, without subsidiary reasoning. If we exclude this distinction and the arbitrary actions in dealing with the most elementary properties of geometric figures that follow from it, the rest of Greek mathematics meets the highest modern standards. With respect to the logical substantiation of the concepts and strictness in deduction it is incomparably higher than European mathematics until the mid-nineteenth century.

However, the way of thinking expressed in Plato's philosophy also had a negative influence. Above all it led to a certain "squeamishness" among the learned Greeks—a lack of desire to work on problems with practical significance. This neglect applied even to approximate calculation. "It is shameful for a free man to study approximate calculations, this is the slave's lot," it was said at that time. Indeed, approximate calculations do not lead to true relationships, and this means that they have no relation to the world of ideas; approximate calculations are on a level with raising olive trees or trading in olive oil. Such a position of course restricted the influx of new problems and ideas and fostered a canonization and regimentation of scientific thought, thus retarding its development. But beyond this, Platonism also had a more concrete negative effect on mathematics. It prevented the Greeks from creating algebraic language. This could be done only by the less educated and more practical Europeans. Later on we shall consider in more detail the history of the creation of modern algebraic language and the inhibiting role of Platonism, but first we shall discuss the answers given by modern science to the questions posed in Platonic times and how the answers given by Plato look in historical retrospect.

■ WHAT IS MATHEMATICS?

FOR US MATHEMATICS is above all a language that makes it possible to create a certain kind of models of reality: mathematical models. As in any other language (or branch of language) the linguistic objects of mathematics, mathematical objects, are material objects that fix definite functional units, mathematical concepts. When we say that the objects "fix functional units" we take this to mean that a person, using the discriminating capabilities of his brain, performs certain linguistic actions on these objects or in relation to them. It is plain that it is not the concrete form (shape, weight, smell) of the mathematical object which is important in mathematics; it is the linguistic activity related to it. Therefore the terms "mathematical object" and "mathematical concept" are often used as synonyms. Linguistic activity in mathematics naturally breaks into two parts: the establishment of a relationship between mathematical objects and nonlinguistic reality (this activity defines the meanings of mathematical concepts), and the formulations of conversions within the language, mathematical calculations and proofs. Often only the second part is what we call "mathematics" while we consider the first as the "application of mathematics."

Points, lines, right triangles, and the like are all mathematical objects. They make up our geometric drawings or stereometric models: spots of color, balls of modeling clay, wires, pieces of cardboard, and the like. The meanings of these objects are known. The point, for example, is an object whose dimensions and shape may be neglected. Thus the "point" is simply an abstract concept which characterizes the relation of an object to its surroundings. In some cases we view our planet as a point. But when we construct a geometric model we usually make a small spot of color on the paper and say, "Let point A be given." This spot of color is in fact linguistic object L_i, and the planet Earth may be the corresponding object (referent) R_i. There are no other "true" or "ideal" points, that is, without dimensions. It is often said that there are no "true" points in nature, but that they exist only in our imagination. This commonplace statement is either absolutely meaningless or false, depending on how it is interpreted. In any case it is harmful, because it obscures the essence of the matter. There are no "true" points in our imagination and there cannot be any. When we say that we are picturing a point we are simply picturing a very small object. Only that which can be made up of the data of sensory experience can be imagined, and by no means all of that. The number 1,000, for example, cannot be imagined. Large numbers, ideal points, and lines exist not in our imagination, but in our language, as linguistic objects we handle in a certain way. The rules for handling them reveal the essence of mathematical concepts, specifically the "ideality of the point." The dimensions of points on a drawing do not influence the development of the proof, and if two points must be set so close that they merge into one, we can increase the scale.

But aren't the assertions of mathematics characterized by absolute precision and correctness which differs sharply from the content of empirical knowledge—which is primarily approximate and hypothetical? We can find by measurement that two segments are approximately equal, but never that they are exactly equal; such assertions are the privilege of mathematics. On the basis of long centuries of human experience we can predict every evening that the sun will rise again the next morning. But this prediction is nothing more than a hypothesis, although an extremely probable one. It is not impossible that somewhere in the interior of the sun or outside it a cosmic catastrophe of unknown nature is coming to a head which will cause the sun to go out or break into parts. But when we say that adding two and two will give four or that the equation $x^2 = 2$ has no rational solutions, we are convinced that these predictions are absolutely correct and will be true always and everywhere, even if the sun and the entire galaxy as well break into little pieces. We simply cannot imagine that it could be otherwise. Consequently, there is a difference between mathematical models of reality and other models which are made up of the content of our everyday experience of the natural sciences. What is the nature of this difference?

■ PRECISION IN COMPARING QUANTITIES

IT IS EASY to see that the absolute precision in comparing measurable objects in mathematics and the absolute definiteness of mathematical assertions are simply results of the fact that mathematical language is a discrete cybernetic system. But is it really discrete? There is no doubt with respect to arithmetic, algebra, and in general the language of symbols. If the top part of the numeral 2 is enlarged or decreased in size it will not become 2.01 or 1.99. A text consisting of N symbols is a cybernetic system of N subsystems, each of which can be pictured as a cell containing a symbol. Suppose that the full number of different symbols is n; then each subsystem may be in one of n states. But geometric language, the language of figures, seems at first glance to be a continuous system. Lines on a drawing may have arbitrary length, form arbitrary angles, and so on. Nonetheless, *in action* geometric language proves to be a discrete system. The details of a geometric drawing, such as the values of the lengths of segments and the magnitudes of angles, play no part in the development of the proof or in decoding the drawing. The only essential things are such characteristics of the drawing as whether two given straight lines intersect, whether a given straight line passes through a given point, whether a given point lies at the intersection of a given straight line and a given circle, and so on. All of this information may be coded in text using some special system of designations or simply in the Russian or English language. The language of geometry can be compared with the language of chess-playing. The chess pieces never occupy the exact centers of the squares of the board and they may even protrude in part outside their own square, but this has no effect at all on the moves the pieces can make.

Assertions of the absolutely exact equality of segments, angles, and the like are simply certain states of the "geometric language" system. Because this system is discrete and deterministic, on the condition that the rules of logical deduction are observed, if it follows from the conditions of the problem that $AB = BC$, we shall invariably receive this result, no matter how many times we repeat the proof (it is assumed, of course, that the system of axioms is noncontradictory; in mathematics only such systems have a right to exist). Because the condition of the problem is already formulated in geometric language, the entire path from the condition to the result is a syntactical conversion $L_1 \rightarrow L_2$ within a discrete linguistic system. The assertions of empirical language have an entirely different status. By itself this language is, of course, discrete also, but empirical assertions reflect semantic conversions $L_1 \rightarrow S_1$ leading us into the area of nonlinguistic activity which is neither discrete nor deterministic. When we say that two rods have equal length this means that every time we measure them the result will be the same. Experience, however, teaches us that if we can increase the precision of measurement without restriction, sooner or later we shall certainly obtain different values for the length, because an empirical assertion of absolutely exact equality is completely senseless. Other assertions of empirical language which have meaning and can be expressed in the language of predicate calculus, for example "rod no. 1 is smaller than rod no. 2," possess the same "absolute precision" (which is a trivial consequence of the discrete nature of the language) as mathematical assertions of the equality of segments. This assertion is either "exactly" true or "exactly" false. Because of variations in the measuring process, however, neither is absolutely reliable.

■ THE RELIABILITY OF MATHEMATICAL ASSERTIONS

NOW LET US DISCUSS the reliability of mathematical assertions. Plato deduced it from the ideal nature of the object of mathematics, from the fact that mathematics does not rely on the illusory and changing data of sensory experience. According to the mathemetician, drawings and symbols are nothing but a subsidiary means for mathematics; the real objects Plato deals with are contained in his imagination and represent the result of perception of the world of ideas through reason, just as sensory experience is the result of perception of the material world through the sense organs. Imagination obviously plays a crucial part in the work of the mathematician (as it does, we might note, in all other areas of creative activity). But it is not entirely correct to say that mathematical objects are *contained* in the imagination; basically they are still contained in drawings and texts, and the imagination takes them up only in small parts. Rather than holding mathematical objects in our imagination we *pass them through* and the characteristics of our imagination determine the functioning of mathematical language. As for the source which determines the content of our imagination, here we disagree fundamentally with Plato. The source is the same sensory experience used in the empirical sciences. Therefore, even though it uses the mediation of imagination, mathematics creates models of the very same, unique (as far as we know) world we live in.

However, although they constructed a stunningly beautiful edifice of logically strict proofs, the Greek mathematicians nonetheless left a number of gaps in the structure; and these gaps, as we have already noticed, lie on the lowest stories of the edifice—in the area of definitions and the most elementary properties of the geometric figures. And this is evidence of a veiled reference to the sensory experience so despised by the Platonists. The mathematics of Plato's times provides even clearer material than does present-day mathematics to refute the thesis that mathematics is independent of experience.

The first statement proved in Euclid's first book contains a method of constructing an equilateral triangle according to a given side. The method is as follows (see figure 10.3). Suppose AB is the given side of the triangle. Taking point A as the center we describe circle π_A with radius AB. We describe a similar circle (π_B) from point B. We use C to designate either of the points of intersection of these circles. Triangle ABC is equilateral, for $AC = CB = AB$.

There is a logical hole in this reasoning: how does it follow that the circles constructed by us will intersect at all? This is a question fraught with complications, for the fact that point of intersection C exists cannot be related either to the attributes of a circle or even to the attributes of a pair of circles (for they by no means always intersect). We are dealing here with a more specific characteristic of the

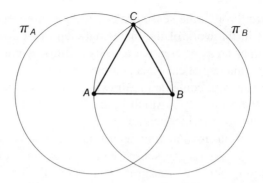

Figure 10.3. Construction of an equilateral triangle.

given situation. Euclid probably sensed the existence of a hole here, but he could not find anything to plug it with.

But how are we certain that circles π_A and π_B intersect? In the last analysis, needless to say, we know from experience. From experience in contemplating and drawing straight lines, circles, and lines in general, from unsuccessful attempts to draw circles π_A and π_B so that they do not intersect.

So Plato's view that the mathematics of his day was entirely independent of experience cannot be considered sound. But the question of the nature of mathematical reliability requires further investigation, for to simply make reference to experience and thus equate mathematical reliability with empirical reliability would mean to rush to the opposite extreme from Platonism. Certainly, we feel clearly that mathematical reliability is somehow different from empirical reliability, but how?

The assertion that circles of radius AB with centers of A and B intersect (for brevity we shall designate this assertion E_I) seems to us almost if not completely reliable; we simply cannot imagine that they would not intersect. *We cannot imagine. . . .* This is how mathematical reliability differs from the empirical! When we are talking about the sun rising tomorrow, we can imagine that the sun will not rise and it is only on the basis of experience that we believe that it probably will rise. Here there are two possibilities and the prediction as to which one will happen is probablistic. But when we say that two

times two is four and that circles constructed as indicated above intersect we cannot imagine that it could be otherwise. We see no other
possibility, and therefore these assertions are perceived as absolutely
reliable and independent of concrete facts we have observed.

■ IN SEARCH OF AXIOMS

IT IS VERY INSTRUCTIVE for an understanding of the nature of mathematical reliability to carry our analysis of the assertion E_I through to
the end. Because we still have certain doubts that the circles in figure
10.3 necessarily intersect, let us attempt to picture a situation where
they do not. If this attempt fails completely it will mean that assertion
E_I is mathematically reliable and cannot be broken down into simpler
assertions; then it should be adopted as an axiom. But if through
greater or lesser effort of imagination we are able to picture a situation in which π_A and π_B do not intersect, it must be expected that this
situation contradicts some simpler and deeper assertions which do
possess mathematical reliability. Then we shall adopt them as axioms
and the existence of the contradiction will serve as proof of E_I. This
is the usual way to establish axioms in mathematics.

First let us draw circle π_A. Then we shall put the point of the
compass at point B and the writing element at point A and begin to
draw circle π_B. We shall move from the center of circle π_A toward
its periphery and at a certain moment (this is how we picture it in our
imagination) we must either intersect circle π_A or somehow skip over
it, thus breaking circle π_B (see figure 10.4). But we imagine circle π_B
as a continuous line and it becomes clear to us that the attributes of
continuousness, which are more fundamental and general than the
other features of this problem, lie at the basis of our confidence that
circles π_A and π_B will intersect. Therefore we set as our goal proving
assertion E_I beginning with the attributes of continuousness of the
circle. For this we shall need certain considerations related to the
order of placement of points on a straight line. We include the concepts of continuousness and order among the basic, undefined concepts of geometry, like the concepts of the point, the straight line, or
distance.

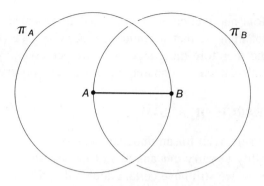

Figure 10.4.

Here is one possible way to our goal. We introduce the concept of "inside" (applicable to a circle) by means of the following definition:

D₁: It is said that point A lies *inside* circle π if it does not lie on π and any straight line passing through point A intersects π at two points in such a way that point A lies between the points of intersection. If the point is neither on nor inside the circle it is said that it lies *outside* the circle.

The concept of "between" characterizes the order of placement of three points on a straight line. It may be adopted as basic and expressed, through the more general concept of "order," by the following definition:

D₂: It is said that point A is located *between* points B_1 and B_2 if these three points are set on one straight line and during movement along this line they are encountered in the order B_1, A, and B_2 or B_2, A, and B_1.

We shall adopt the following propositions as axioms:

A₁: The center of a circle lies inside it.

A₂: The arc of a circle connecting any two points of the circle is continuous.

A₃: If point A lies inside circle π and point B is outside it, and these two points are joined by a continuous line, then there is a point where this line intersects the circle.

Relying on these axioms, let us begin with the proof. According to the statement of the problem, circle π_B passes through center A of circle π_A. If we have confidence that there is at least one point of circle π_B that does not lie inside π_A we shall prove E_1. Indeed, if it lies on π_A then E_1 has been proved. If it lies outside π_A then the arc of circle π_B connects it with the center, that is, with an inside point of circle π_A. Therefore, according to axioms A_2 and A_3 there is a point of intersection of π_B and π_A.

But can we be confident that there is a point on circle π_B which is outside π_A? Let us try to imagine the opposite case. It is shown in figure 10.5. This is the second attempt to imagine a situation which contradicts the assertion being proved. Whereas the first attempt immediately came into explicit contradiction with the continuousness of a circle, the second is more successful. Indeed, stretching things a bit we can picture this case. We take a compass, put its point at point B and the pencil at point A. We begin to draw the circle without taking the pencil from the paper and when the pencil returns to the starting point of the line we remove it and see that we have figure 10.5. And why not?

To prove that this is impossible we must prove that in this case the center of circle π_B is necessarily outside it. We shall be helped in this by the following theorem:

T_1: If circle π_1 lies entirely inside circle π_2 then every inside point of circle π_1 is also an inside point of circle π_2.

To prove this we shall take an arbitrary inside point A of circle

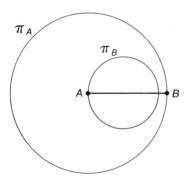

Figure 10.5.

π_1 (see figure 10.6). We draw a straight line through it. According to definition $\mathbf{D}_1$ it intersects π_1 at two points: B_1 and B_2. Because B_1 (just as B_2) lies inside π_2 this straight line also intersects π_2 at two points: C_1 and C_2. We have received five points on a straight line and they are connected by the following relationships of order: A lies between B_1 and B_2; B_1 and B_2 lie between C_1 and C_2. That point A proves to be between points C_1 and C_2 in this situation seems so obvious to us that we shall boldly formulate it as still another axiom.

$\mathbf{A}_3$: If points B_1 and B_2 on a straight line both lie between C_1 and C_2, then any point A lying between B_1 and B_2 also lies between C_1 and C_2.

Because we can take any point inside π_1 as A and we can draw any straight line through it, theorem $\mathbf{T}_1$ is proven.

Now it is easy to complete the proof of E_1. If circle π_B lies entirely inside π_A then according to theorem T_1 its center B must also lie inside π_A. But according to the statement of the problem point B is located on π_A. Therefore π_B contains at least one point which is not inside in relation to π_A.

So to prove one assertion E_1 we needed four assertions (axioms $\mathbf{A}_1$–$\mathbf{A}_4$), but then these assertions express very fundamental and general models of reality related to the concepts of continuousness and order and we cannot even imagine that they are false. The only question that can be raised refers to axiom $\mathbf{A}_1$ which links the concept of center, which is metrical (that is, including the concept of measure-

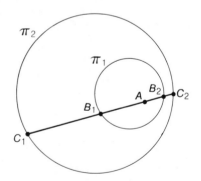

Figure 10.6.

ment) in nature, with the concept of "inside," which relies exclusively on the concepts of continuousness and order. It may be desired that this connection be made using simpler geometric objects, under conditions which are easier for the functioning of imagination. This desire is easily met. For axiom A_1 let us substitute the following axiom:

A_1': if on a straight line point A and a certain distance (segment) R are given, then there are exactly two points on the straight line which are set at distance R from point A, and point A lies between these two points.

Relying on this axiom we shall prove assertion A_1 as a theorem. We shall draw an arbitrary straight line through the center of the circle. According to axiom A_1' there will be two points on it which are set at distance R (radius of the circle) from the center. Because a circle is defined as the set of all points which are located at distance R from the center, these points belong to the circle. According to axiom A_1' the center point lies between them and therefore, according to definition D_1, it is an inside point. In this way axiom A_1 has been reduced to axiom A_1'. Now try to imagine a point on a straight line which does not have two points set on different sides from it at the given distance!

■ CONCERNING THE AXIOMS OF ARITHMETIC AND LOGIC

THE PRIMARY PROPOSITIONS of arithmetic in principle possess the same nature as the primary propositions of geometry, but they are perhaps even simpler and more obvious and denial of them is even more inconceivable than denial of geometric axioms. As an example let us take the axiom which says that for any number

$$a + 0 = a$$

The number 0 depicts an empty set. Can you imagine that the number of elements in some certain set would change if it were united with an empty set? Here is another arithmetic axiom: for any numbers a and b

$$a + (b + 1) = (a + b) + 1,$$

that is, if we increase the number b by one and add the result to a, we shall obtain the same number as if we were to add a and b first and then increase the result by one. If we analyze why we are unable to imagine a situation that contradicts this assertion, we shall see that it is a matter of the same considerations of continuousness that also manifest themselves in geometric axioms. In the process of counting, it is as if we draw continuous lines connecting the objects being counted with the elements of a standard set and, of course, lines in time (let us recall the origin of the concept "object") whose continuousness ensures that the number is identical to itself.

Natural auditory language transferred to paper gives rise to linear language, that is, a system whose subsystems are all linear sequences of signs. Signs are objects concerning which it is assumed only that we are able to distinguish identical ones from different ones. The linearity of natural languages is a result of the fact that auditory language unfolds in time and the relation of following in time can be modeled easily by the relation of order of placement on a timeline. The specialization of natural language led to the creation of the linear, symbolic mathematical language which now forms the basis of mathematics.

Operating within the framework of linear symbolic languages we are constantly taking advantage of certain other attributes which seem so obvious and self-evident that we don't even want to formulate them in the form of axioms. As an example let us take this assertion: if symbol a is written to the left of symbol b and symbol c is written on the right the same word (sequence of characters) will be received as when b is written to the right of a and followed by c. This assertion and others like it possess mathematical reliability for we cannot imagine that it would be otherwise. One of the fields of modern mathematics, the theory of semi-groups, studies the properties of linear symbolic systems from an axiomatic point of view and declares the simplest of these properties to be axioms.

All three kinds of axioms, geometric, arithmetic, and linear-symbolic, possess the same nature and in actuality rely on the same fundamental concepts, concepts such as identity, motion, continuousness, and order. There is no difference in principle among these groups of axioms. And if one term were to be selected for them they should be called geometric or geometric-kinematic because they all reflect the attributes of our space-time experience and space-time imagination. The only more or less significant difference which can be found is in the group of "properly geometric" axioms; some of the axioms concerning straight lines and planes reflect more specific experience related to the existence of solid bodies. The same thing evidently applies to metric concepts. But this difference too is quite arbitrary. Can we say anything serious about those concepts which we would have if there were no solid bodies in the world?

Thus far we have been discussing the absolute reliability of axioms. But where do we get our confidence in the reliability of assertions obtained by logical deduction from axioms? From the same source. Our imagination refuses to permit a situation in which by logical deduction we obtain incorrect results from correct premises. Logical deduction consists of successive steps. At each step, relying on the preceding proposition, we obtain a new one. From a review of formal logical deduction (chapter 11), it will be seen that our confidence that at every step we can only receive a true proposition from higher true propositions is based on logical axioms [2] which seem to us just as reliable as the mathematical axioms considered above. And this is for the same reason, that the opposite situation is absolutely inconceivable. Having this confidence we acquire confidence that no matter how many steps a logical deduction may contain it will still possess this attribute. Here we are using the following very important axiom:

The axiom of induction: Let us suppose that function $f(x)$ leaves attribute $P(x)$ unchanged, that is

$$(\forall x)\{P(x) \supset P[f(x)]\}.$$

---

[2] For those who are familiar with mathematical logic let us note that this is in the broad sense, including the rules of inference.

We will use $f^n(x)$ to signify the result of sequential n-time application of function $f(x)$, that is

$$f^1(x) = f(x), f^n(x) = f[f^{n-1}(x)].$$

Then $f^n(x)$ will also leave attribute P(x) unchanged for any n, that is

$$(\forall n)(\forall x)\{P(x) \supset P[f^n(x)]\}.$$

By their origin and nature logical axioms and the axiom of induction (which is classed with arithmetic because it includes the concept of number) do not differ in any way from the other axioms; they are all mathematical axioms. The only difference is in how they are used. When mathematical axioms are applied to mathematical assertions they become elements of a metasystem within the framework of a system of mathematically reliable assertions and we call them logical axioms. Thanks to this, the system of mathematically reliable assertions becomes capable of development. The great discovery of the Greeks was that it is possible to add one certainty to another certainty and thus obtain a new certainty.

■ DEEP-SEATED PILINGS

THE DESCRIPTION of mathematical axioms as models of reality which are true not only in the sphere of real experience but also in the sphere of imagination relies on their subjective perception. Can it be given a more objective characterization?

Imagination emerges in a certain stage of development of the nervous system as arbitrary associating of representations. The preceding stage was the stage of nonarbitrary associating (the level of the dog). It is natural to assume that the transition from nonarbitrary to arbitrary associating did not produce a fundamental change in the material at the disposal of the associating system, that is, in the representations which form the associations. This follows from the hierarchical principle of the organization and development of the nervous system in which the superstructure of the top layers has a weak influence on the lower ones. And it follows from the same principle that in the process of the preceding transition, from fixed concepts to nonarbitrary associating, the lowest levels of the system of concepts remained unchanged and conditioned those universal, deep-seated properties of representations that were present before associating and that associating could not change. Imagination cannot change them either. These properties are invariant in relation to the transformations made by imagination. And they are what mathematical axioms rely on.

If we picture the activity of the imagination as shuffling and fixing certain elements, "pieces" of sensory perception, then axioms are models which are true for any piece and, therefore, for any combination of them. The ability of the imagination to break sensory experience up into pieces is not unlimited; emerging at a certain stage of development it takes the already existing system of concepts as its background, as a foundation not subject to modification. Such profound concepts as motion, identity, and continuousness were part of this background and therefore the models which rely on these concepts are universally true not only for real experience but also for any construction the imagination is capable of creating.

Mathematics forms the frame of the edifice of natural sciences. Its axioms are the support piles that drive deep into the neuronal concepts, below the level where imagination begins to rule. This is the reason for the stability of foundation which distinguishes mathematics from empirical knowledge. Mathematics ignores the superficial associations which make up our everyday experience, preferring to continue constructing the skeleton of the system of concepts which was begun by nature and set at the lowest levels of the hierarchy. And this is the skeleton on which the "noncompulsory" models we class with the natural sciences will form, just as the "noncompulsory" associations of representations which make up the content of everyday experience form on the basis of inborn and "compulsory" concepts of the lowest level. The requirements dictated by mathematics are compulsory; when we are constructing models of reality we cannot bypass them even if we want to. Therefore we always refer the possible falsehood of a theory beyond the sphere of mathematics. If a discrepancy is found between the theory and the experiment it is the external, "noncompulsory" part of the theory that is changed, but no one would ever think of expressing the assumption that, in such a case, the equality $2 + 2 = 4$ has proved untrue.

The "compulsory" character of classical mathematical models does not contradict the appearance of mathematical and physical theories which at first glance conflict with our space-time intuition (for example, non-Euclidian geometry or quantum mechanics). These theories are linguistic models of reality whose usefulness is seen not in the sphere of everyday experience but in highly specialized situations. They do not destroy and replace the classical models; they continue them. Quantum mechanics, for example, relies on classical mechanics. And what theory can get along without arithmetic? The paradoxes and contradictions arise when we forget that the concept-constructs which are included in a new theory are new concepts, even when they are given old names. We speak of a "straight line" in non-Euclidian geometry and call an electron a "particle" although the linguistic activity related to these words (proof of theorems and quantum mechanics computations) is not at all identical to that for the former theories from which the terms were borrowed. If two times two is not four then either two is not two, times is not times, or four is not four.

The special role of mathematics in the process of cognition can be expressed in the form of an assertion, that mathematical concepts and axioms are not the result of cognition of reality, rather they are a condition and form of cognition. This idea was elaborated by Kant and we may agree with it if we consider the human being to be entirely given and do not ask why these conditions and forms of cognition are characteristic of the human being. But when we have asked this question we must reach the conclusion that they themselves are models of reality developed in the process of evolution (which, in one of its important aspects, is simply the process of cognition of the world by living structures). From the point of view of the laws of nature there is no fundamental difference between mathematical and empirical models; this distinction reflects only the existence in organization of the human mind of a certain border line which separates inborn models from acquired ones. The position of this line, one must suppose, contains an element of historical accident. If it had originated at another level, perhaps we would not be able to imagine that the sun may fail to rise or that human beings could soar above the earth in defiance or gravity.

■ PLATONISM IN RETROSPECT

PLATO'S IDEALISM was the result of a sort of projection of the elements of language onto reality. Plato's "ideas" have the same origin as the spirits in primitive thinking; they are the imagined meanings of really existing names. In the first stages of the development of critical thinking the nature of abstraction in the interrelationship of linguistic objects and nonlinguistic activity is not yet correctly understood. The primitive name-meaning unit is still pressing on people an idea of a one-to-one correspondence between names and their meanings. For words that refer to concrete objects the one-to-one correspondence seems to occur because we picture the object as some one thing. But what will happen with general concepts (universals)? In the sphere of the concrete there is no place at all for their meanings; everything has been taken up by "unique" concepts, for a label with a name can be attached to each object. The empty place that forms is filled by the "idea." Let us emphasize that Plato's idealism is far from including an assertion of the primacy of the spiritual over the material, which is to say it is not *spiritualism* (this term, which is widely used in Western literature, is little used in our country and is often replaced by the term "idealism," which leads to inaccuracy). According to Plato spiritual experience is just as empirical as sensory experience and it has no relation to the world of ideas. Plato's "ideas" are pure specters, and they are specters born of sensory, not spiritual, experience.

From a modern cybernetic point of view only a strictly defined, unique situation can be considered a unique concept. This requires an indication of the state of all receptors that form the input of the nervous system. It is obvious that subjectively we are totally unaware of concepts that are unique in this sense. Situations that are merely similar become indistinguishable somewhere in the very early stages of information processing and the representations with which our consciousness is dealing are generalized states, that is to say, general, or abstract, concepts (sets of situations). The concepts of definite objects which traditional logic naively takes for the primary elements of sensory experience and calls "unique" concepts are in reality, as was shown above, very complex constructions which require analysis of the moving picture of situations and which rely on more elementary abstract concepts such as continuousness, shape, color, or spatial relations. And the more "specific" a concept is from the logical point of view, the more complex it will be from the cybernetic point of view. Thus, a specific cat differs from the abstract cat in that a longer moving picture of situations is required to give meaning to the first concept than to the second. Strictly speaking the film may even be endless, for when we have a specific cat in mind we have in mind not only its "personal file" which has been kept since its birth, but also its entire genealogy. There is no fundamental difference in the nature of concrete and abstract concepts; they both reflect characteristics of the real world. If there is a difference, it is the opposite of what traditional logic discerns: abstract, general concepts of sensory and spiritual experience (which should not be confused with mathematical *constructs*) are simpler and closer to nature than concrete concepts which refer to the definite objects. Logicians were confused by the fact that concrete concepts appeared in language earlier than abstract ones did. But this is evidence of their relatively higher position in the hierarchy of neuronal concepts, thanks to which they emerged at the point of connection with linguistic concepts.

The Platonic theory of ideas, postulating a contrived, ideal existence of generalized objects, puts one-place predicates (attributes) in a position separate from multiplace predicates (relations). This theory assigned attributes the status of true existence but denied it to relations, which became perfectly evident in Aristotle's logic. The concrete, visual orientation and static quality in thinking which were so characteristic of the Greeks in the classical period came from this. In the next chapter we shall see how this way of thinking was reflected in the development of mathematics.

CHAPTER ELEVEN
From Euclid to Descartes

■ NUMBER AND QUANTITY

DURING THE TIME of Pythagoras and the early Pythagoreans, the concept of number occupied the dominant place in Greek mathematics. The Pythagoreans believed that God had made numbers the basis of the world order. God is unity and the world is plurality. The divine harmony in the organization of the cosmos is seen in the form of numerical relationships. A substantial part in this conviction was played by the Pythagoreans' discovery of the fact that combinations of sounds which are pleasant to hear are created in the cases where a string is shortened by the ratios formed by whole numbers such as 1:2 (octave), 2:3 (fifth), 3:4 (fourth), and so on. The numerical mysticism of the Pythagoreans reflected their belief in the fact that, in the last analysis, all the uniformities of natural phenomena derive from the properties of whole numbers.

We see here an instance of the human inclination to overestimate new discoveries. The physicists of the late nineteenth century, like the Pythagoreans, believed that they had a universal key to all the phenomena of nature and with proper effort would be able to use this key to reveal the secret of any phenomenon. This key was the notion that space was filled by particles and fields governed by the equations of Newton and Maxwell. With the discovery of radioactivity and the diffraction of electrons, however, the physicists' arrogant posture crumbled.

In the case of the Pythagoreans the same function was performed by discovery of the existence of incommensurable line segments, that is, segments such that the ratio of their lengths is not expressed by any ratio of whole numbers (rational number). The side of a square and its diagonal are incommensurable, for example. It is easy to prove this statement using the Pythagorean theorem. In fact, let us suppose the opposite, namely that the diagonal of a square stands in some ratio $m:n$ to its side. If the numbers m and n have common factors they can be reduced, so we shall consider that m and n do not have common factors. This means that in measuring length by some unitary segment, the length of the side is n and the length of the diagonal is m. It follows from the Pythagorean theorem that the equality $m^2 = 2n^2$ must occur. Therefore, m^2 must be divisible by 2, and consequently 2 must be among the factors of m, that is, $m = 2m_1$. Making this substitution we obtain $4m_1^2 = 2n^2$, that is, $2m_1^2 = n^2$. This means that n also must be divisible by 2, which contradicts the assumption that m and n do not have common factors. Aristotle often refers to this proof. It is believed that the proof had already been discovered by the Pythagoreans.

If there are quantities which for a given scale are not expressed by numbers then the number can no longer be considered the foundation of foundations; it is removed from its pedestal. Mathematicians then must use the more general concept of *geometric quantity* and study the relations among quantities that may (although only occasionally) be expressed in a ratio of whole numbers. This approach lies at the foundation of all Greek mathematics beginning with the classical period. The relations we know as algebraic equalities were known to the Greeks in geometric formulation as relations among lengths, areas, and volumes of figures constructed in a definite manner.

■ GEOMETRIC ALGEBRA

FIGURE 11.1 shows the well-known geometric interpretation of the relationship $(a+b)^2 = a^2 + 2ab + b^2$. The equality $(a+b)(a-b) = a^2 - b^2$, which is equally commonplace from an algebraic point of

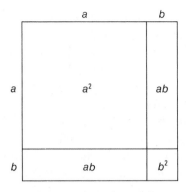

Figure 11.1. Geometric interpretation of the identity

$$(a + b)^2 = a^2 + 2ab + b^2.$$

view, requires more complex geometric consideration. The following theorem from the second book of Euclid's *Elements* corresponds to it (see figure 11.2).

"If a straight line be cut into equal and unequal segments, the rectangle contained by the unequal segments of the whole together with the square on the straight line between the points of the section is equal to the square on the half." [1]

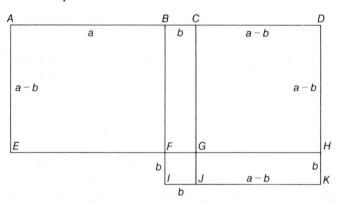

Figure 11.2. Geometric interpretation of the identity

$$(a - b)(a + b) = a^2 - b^2.$$

[1] *The Thirteen Books of Euclid's Elements*, translated and annotated by T. L. Heath, (Cambridge: Cambridge University Press, 1908), vol. 1, p. 382.

The theorem is proved as follows. Rectangle *ABFE* is equal to rectangle *BDHF*. Rectangle *BCGF* is equal to rectangle *GHKJ*. If square *FGJI* is added to these two rectangles (which together form rectangle *ACGE*, which is "contained by the unequal segments of the whole") what we end up with is precisely rectangle *BDKI*, which is constructed "on the half." Thus we have the equality $(a + b)$ $(a - b) + b^2 = a^2$, which is equivalent to the equality above but does not contain the difficult-to-interpret subtraction of areas.

Clearly, if these very simple algebraic relations require great effort to understand the formulation of the theorem—as well as inventiveness in constructing the proof—when they are expressed geometrically, then it is impossible to go far down this path. The Greeks proved themselves great masters in everything concerning geometry proper, but the line of mathematical development that began with algebra and later gave rise to the infinitesimal analysis and to modern axiomatic theories (that is to say, the line of development involving the use of the language of symbols rather than the language of figures) was completely inaccessible to them. Greek mathematics remained limited, confined to the narrow framework of concepts having graphic geometric meaning.

■ ARCHIMEDES AND APOLLONIUS

DURING THE ALEXANDRIAN EPOCH (330–200 B.C.) two great learned men lived in whose work Greek mathematics reached its highest point. They were Archimedes (287–212 B.C.) and Apollonius (265?–170? B.C.)

In his works on geometry Archimedes goes far beyond the limits of the figures formed by straight lines and circles. He elaborates the theory of conic sections and studies spirals. Archimedes's main achievement in geometry is his many theories on the areas, volumes, and centers of gravity of figures and bodies formed by other than just straight lines and plane surfaces. He uses the "method of exhaustion." To illustrate the range of problems solved by Archimedes we shall list the problems included in his treatise entitled *The Method* whose purpose, as can be seen from the title, is not a full summary of results but rather an explanation of the method of his work. *The Method* contains solutions to the following 13 problems: area of a parabolic segment, volume of a sphere, volume of a spheroid (ellipsoid of rotation), volume of a segment of a paraboloid of rotation, center of gravity of a segment of a paraboloid of rotation, center of gravity of a hemisphere, volume of a segment of a sphere, volume of a segment of a spheroid, center of gravity of a segment of a sphere, center of gravity of a segment of a spheroid, center of gravity of a segment of a hyperboloid of rotation, volume of a segment of a cylinder, and volume of the intersection of two cylinders (the last problem is without proof).

Archimedes's investigations in the field of mechanics were just as important as his work on geometry. He discovered his famous "law" and studied the laws of equilibrium of bodies. He was extraordinarily skillful in making different mechanical devices and attachments. It was thanks to machines built under his direction that the inhabitants of Syracuse, his native city, repulsed the Romans' first attack on their city. Archimedes often used mechanical arguments as support in deriving geometric theorems. It would be a mistake to suppose, however, that Archimedes deviated at all from the traditional Greek way of thinking. He considered a problem solved only when he had found a logically flawless geometric proof. He viewed his mechanical inventions as amusements or as practical concerns of no scientific importance whatsoever. "Although these inventions," Plutarch writes, "made his superhuman wisdom famous, he nonetheless wrote nothing on these matters because he felt that the construction of all machines and all devices for practical use in general was a low and ignoble business. He himself strove only to remove himself, by his handsomeness and perfection, far from the kingdom of necessity."

Of all his achievements Archimedes himself was proudest of his proof that the volume of a sphere inscribed in a cylinder is two thirds of the volume of the cylinder. In his will he asked that a cylinder with an inscribed sphere be shown on his gravestone. After the

Romans took Syracuse and one of his soldiers (against orders, it is said) killed Archimedes, the Roman general Marcellus authorized Archimedes' relatives to carry out the wish ofthe deceased.

Apollonius was primarily famous for his work on the theory of conic sections. His work is in fact a consistent algebraic investigation of second-order curves, expressed in geometric language. In our day any college student can easily repeat Appolonius's results by employing the methods of analytic geometry. But Apollonius needed to show miraculous mathematical intuition and inventiveness to do the same thing within a purely geometric approach.

■ THE DECLINE OF GREEK MATHEMATICS

"AFTER APOLLONIUS," writes B. van der Waerden, "Greek mathematics comes to a dead stop. It is true that there were some epigones, such as Diocles and Zenodorus, who, now and then, solved some small problem, which Archimedes and Apollonius had left for them, crumbs from the board of the great. It is also true that compendia were written, such as that of Pappus of Alexandria (300 A.D.); and geometry was applied to practical and to astronomical problems, which led to the development of plane and spherical trigonometry. But apart from trigonometry, nothing great nothing new appeared. The geometry of the conics remained in the form Apollonius gave it, until Descartes. Indeed the works of Apollonius were but little read and were even partly lost. The *Method* of Archimedes was lost sight of and the problem of integration remained where it was, until it was attacked new in the seventeenth century. . . ." [2]

The decline of Greek mathematics was in part caused by external factors—the political storms that engulfed Mediterranean civilization. Nonetheless, internal factors were decisive. In astronomy, van der Waerden notes, development continued steadily along an ascending line; there were short and long periods of stagnation, but after them the work was taken up again from the place where it had

[2] B. van der Waerden, *Science Awakening* (New York: Oxford University Press, 1969), p. 264.

stopped. In geometry, however, regression plainly occurred. The reason is found, of course, in the lack of an algebraic language.

"Equations of the first and second degree," we read in van der Waerden, "can be expressed clearly in the language of geometric algebra and, if necessary, also those of the third degree. But to get beyond this point, one has to have recourse to the bothersome tool of proportions.

"Hippocrates, for instance, reduced the cubic equation $x^3 = V$ to the proportion

$$a : x = x : y = y : b,$$

and Archimedes wrote the cubic

$$x^2 (a - x) = bc^2$$

in the form

$$(a - x) : b = c^2 : x^2."$$

"In this manner one can get to equations of the fourth degree; examples can be found in Apollonius. . . . But one cannot get any further; besides, one has to be a mathematician of genius, thoroughly versed in transforming proportions with the aid of geometric figures, to obtain results by this extremely cumbersome method. Anyone can use our algebraic notation, but only a gifted mathematician can deal with the Greek theory of proportions and with geometric algebra.

Something has to be added, that is, the difficulty of the written tradition.

Reading a proof in Apollonius requires extended and concentrated study. Instead of a concise algebraic formula, one finds a long sentence, in which each line segment is indicated by two letters which have to be located in the figure. To understand the line of thought, one is compelled to transcribe these sentences in modern concise formulas. . . .

An oral explanation makes it possible to indicate the line segments with the fingers; one can emphasize essentials and point out how the proof was found. All of this disappears in the written formulation of the strictly classical style. The proofs are logically sound, but they are not suggestive. One feels caught as in a logical mouse trap, but one fails to see the guiding line of thought.

As long as there was no interruption, as long as each generation could hand over its method to the next, everything went well and the science flourished. But as soon as some external cause brought about an interruption in the oral tradition, and only books remained, it became extremely difficult to assimilate the work of the great precursors and next to impossible to pass beyond it.[3]

But why, despite their high mathematical sophistication and abundance of talented mathematicians, did the Greeks fail to create an algebraic language? The usual answer is that their high mathematical sophistication was in fact what hindered them, more specifically their extremely rigorous requirements for logical strictness in a theory, for the Greeks could not consider irrational numbers, in which the values of geometric quantities are ordinarily expressed, as numbers; if line segments were incommensurate it was considered that a numerical relationship for them simply did not exist. Although this explanation is true in general, still it must be recognized as imprecise and supeficial. Striving toward logical strictness cannot by itself be a negative factor in the development of mathematics. If it acts as a negative factor this will evidently be only in combination with certain other factors and the decisive role in this combination should certainly not be ascribed to the striving for strictness. Perfect logical strictness in his final formulations and proofs did not prevent Archimedes from using guiding considerations which were not strict. Then why did it obstruct the creation of an algebraic language? Of course, this is not simply a matter of a high standard of logical strictness, it concerns the whole way of thinking, the philosophy of mathematics. In creating the modern algebraic language Descartes went beyond the Greek canon, but this in no way means that he sinned against the laws of logic or that he neglected proof. He considered irrational numbers to be "precise" also, not mere substitutions for their approximate values. Some problems with logic arose after the time of Descartes, during the age of swift development of the infini-

---

[3] van der Waerden, *Science Awakening*, p. 266.

tesimal analysis. At that time mathematicians were so carried away by the rush of discoveries that they simply were not interested in logical subtleties. In the nineteenth century came time to pause and think, and then a solid logical basis was established for the analysis.

We shall grasp the causes of the limitations of Greek mathematics after we review the substance of the revolution in mathematics made by Descartes.

■ ARITHMETIC ALGEBRA

ADVANCES IN GEOMETRY forced the art of solving equations into the background. But this art continued to develop and gave rise to *arithmetic algebra*. The emergence of algebra from arithmetic was a typical metasystem transition. When an equation must be solved, whether it is formulated in everyday conversational language or in a specialized language, this is an arithmetic problem. And when the general method of solution is pointed out—by example, as is done in elementary school, or even in the form of a formula—we still do not go beyond arithmetic. Algebra begins when the equations themselves become the object of activity, when the properties of equations and rules for converting them are studied. Probably everyone who remembers his first acquaintance with algebra in school (if this was at the level of understanding, of course, not blind memorization) also remembers the happy feeling of surprise experienced when it turns out that various types of arithmetic problems whose solutions had seemed completely unrelated to one another are solved by the same conversions of equations according to a few simple and understandable rules. All the methods known previously fall into place in a harmonious system, new methods open up, new equations and whole classes of equations come under consideration (the law of branching growth of the penultimate level), and new concepts appear which have absolutely no meaning within arithmetic proper: negative, irrational, and imaginary numbers.

In principle the creation of a specialized language is not essential for the development of algebra. In fact, however, only with the creation of a specialized language does the metasystem transition in people's minds conclude. The specialized language makes it possible to see with one's eyes that we are dealing with some new reality, in this case with equations, which can be viewed as an object of computations just as the objects (numbers) of the preceding level were. People typically do not notice the air they breathe and the language they speak. But a newly created specialized language goes outside the sphere of natural language and is in part nonlinguistic activity. This facilitates the metasystem transition. Of course, the practical advantages of using the specialized language also play an enormous part here; among them are making expressions visible, reducing time spent recopying, and so on.

The Arab scholar Muhammed ibn Musa al-Khwarizmi (780–850) wrote several treatises on mathematics which were translated into Latin in the twelfth century and served as the most important textbooks in Europe for four centuries. One of them, the *Arithmetic*, gave Europeans the decimal system of numbers and the rules (algorithms—the name is based on al-Khwarizmi) for performing the four arithmetic operations on numbers written in this sytem. Another work was entitled *Book of Al Jabr Wa'l Muqabala*. The purpose of the book was to teach the art of solving equations, an art which is essential, as the author writes, "in cases of inheritance, division of property, trade, in all business relationships, as well as when measuring land, laying canals, making geometric computations, and in other cases. . . ." *Al Jabr* and *al Mugabala* are two methods al-Khwarizmi uses to solve equations. He did not think up these methods himself; they were described and used in the *Arithmetica* of the Greek mathematician Diophantus (third century A.D.), who was famous for his methods of solving whole-number ("diophantine") equations. In the same *Arithmetica* of Diophantus we find the rudiments of letter symbolism. Therefore, if anyone is to be considered the progenitor of arithmetic algebra it should obviously be Diophantus. But Europeans first heard of algebraic methods from al-Khwarizmi while the works of Diophantus became known much later. There is no special algebraic symbolism in al-Khwarizmi, not even in rudimentary form. The equations are written in natural language. But for brevity's sake, we shall describe these methods and give our examples using modern symbolic notation.

Al Jabr involves moving elements being subtracted from one part of the equation to the other; al Muqabala involves subtracting the same element from both parts of the equation. Al-Khwarizmi considers these procedures different because he does not have the concept of a negative number.

For example let us take the equation

$$7x - 11 = 5x - 3.$$

Applying the al Jabr method twice, for the 11 and 3, which are to be subtracted, we receive:

$$7x + 3 = 5x + 11.$$

Now we use the al Muqabala method twice, for 3 and 5x. We receive

$$2x = 8.$$

From this we see that

$$x = 4.$$

So although al-Khwarizmi does not use a special algebraic language, his book contains the first outlines of the algebraic approach. Europeans recognized the merits of this approach and developed it further. The very word algebra comes from the name of the first of al-Khawarizmi's methods.

■ ITALY, SIXTEENTH CENTURY

IN THE FIRST HALF of the sixteenth century the efforts of Italian mathematicians led to major changes in algebra which were associated with very dramatic events. Scipione del Ferro (1465–1526), a professor at the University of Bologna, found a general solution to the cubic equation $x^3 + px = q$ where p and q are positive. But del Ferro kept it secret, because it was very valuable in the problem-solving competitions which were held in Italy at that time. Before his death he revealed his secret to his student Fiore. In 1535 Fiore challenged the brilliant mathematician Niccolo Tartaglia (1499–1557) to a contest. Tartaglia knew that Fiore possessed a method of solving the cubic equation, so he made an all-out effort and found the solution himself. Tartaglia won the contest, but he also kept his discovery secret. Finally, Girolamo Cardano (1501–1576) tried in vain to find the algorithm for solving the cubic equation. In 1539 he finally appealed to Tartaglia to tell him the secret. Having received a "sacred oath" of silence from Cardano, Tartaglia unveiled the secret, but only partially and in a rather unintelligible form. Cardano was not satisfied and made efforts to familiarize himself with the manuscript of the late del Ferro. In this he was successful, and in 1545 he published a book in which he reported his algorithm, which reduces the solution of a cubic equation to radicals (the "Cardano formula"). This same book contained one more discovery made by Cardano's student Luigi (Lodovico) Ferrari (1522–1565): the solution of a quartic equation in radicals. Tartaglia accused Cardano of violating his oath and began a bitter and lengthy polemic. It was under such conditions that modern mathematics made its first significant advances.

Using a tool suggests ways to improve it. While striving toward a uniform solution to equations, mathematicians found that it was extremely useful in achieving this goal to introduce certain new objects and treat them as if they were numbers. And in fact they were called numbers, although it was understood that they differed from "real" numbers; this was seen in the fact that they were given such epithets as "false," "fictitious," "incomprehensible," and "imaginary." What they correspond to in reality remained somewhat or entirely unclear. Whether their use was correct also remained debatable. Nonetheless, they began to be used increasingly widely, because with them it was possible to obtain finite results containing only "real" numbers which could not be obtained otherwise. A person consistently following the teachings of Plato could not use "unreal" numbers. But the Indian, Arabic, and Italian mathematicians were by no means consistent Platonists. For them a healthy curiosity and pragmatic considerations outweighed theoretical prohibitions. In this, however, they did make reservations and appeared to be apologizing for their "incorrect" behavior.

All "unreal" numbers are products of the reverse movement in the arithmetic model; formally they are solutions to equations that cannot have solutions in the area of "real" numbers. First of all we must mention negative numbers. They are found in quite developed form in the Indian mathematician Bhascara (twelfth century), who performed all four arithmetic operations on such numbers. The interpretation of the negative number as a debt was known to the Indians as early as the seventh century. In formulating the rules of operations on negative numbers, Bhascara calls them "debts," and calls positive numbers "property." He does not choose to declare the negative number an abstract concept like the positive number. "People do not approve of abstract negative numbers," Bhascara writes. The attitude toward negative numbers in Europe in the fifteenth and sixteenth centuries was similar. In geometric interpretation negative roots are called "false" as distinguished from the "true" positive roots. The modern interpretation of negative numbers as points lying to the left of the zero point did not appear until Descartes' *Géométrie* (1637). Following tradition, Descartes called negative roots false.

Formal operations on roots of numbers that cannot be extracted exactly go back to deep antiquity, when the concept of incommensurability of line segments had not yet appeared. In the fifteenth and sixteenth centuries people handled them cavalierly; they were helped here, of course, by the simple geometric interpretation. An understanding of the theoretical difficulty which arises from the incommensurability of line segments can be seen by the fact that the numbers were called "irrational."

The square of any number is positive; therefore the square root of a negative number does not exist among positive, negative, rational, or irrational numbers. But Cardano was daring enough to use (not without reservations) the roots of negative numbers. "Imaginary" numbers thus appeared. The logic of using algebraic language drew mathematicians inexorably down an unknown path. It seemed wrong and mysterious, but intuition suggested that all these impossible numbers were profoundly meaningful and that the new path would prove useful. And it certainly did.

■ LETTER SYMBOLISM

THE RUDIMENTS of algebraic letter symbolism are first encountered, as mentioned above, in Diophantus. Diophantus used a character resembling the Latin *s* to designate an unknown. It is hypothesized that this designation originates from the last letter of the Greek word for number: ἀριθμόσ (*arithmos*). He also had abbreviated notations for the square, cube, and other degrees of the unknown quantity. He did not have an addition sign; quantities being added were written in a series. Something like an upside-down Greek letter ψ was used as the subtraction sign, while the first letter of the Greek word ισοσ for "equal" was used as the equal sign. Everything else was expressed in words. Known quantities were always written in concrete numerical form, while there were no designations for known, but arbitrary numbers.

Diophantus' *Arithmetica* became known in Europe in 1463. In the late fifteenth and early sixteenth centuries European mathematicians, first Italians and then others, began to use abbreviated notations. These abbreviations gradually wandered from arithmetic algebra to geometric, and unknown geometric quantities also began to be designated by letters. In the late sixteenth century the Frenchman François Vieta (1540–1603) took the next important step. He introduced letter designations for known quantities and was thus able to write equations in general form. Vieta also introduced the term "coefficient." In external appearance Vieta's symbols are still rather far from modern ones. For example, Vieta writes

$$D \text{ in} \frac{\left[\begin{array}{l} B \text{ cubum } 2 \\ -D \text{ cubo} \end{array}\right.}{\begin{array}{c} B \text{ cubo} \\ +D \text{ cubo} \end{array}}$$

instead of our notation

$$\frac{D(2B^3 - D^3)}{B^3 + D^3}$$

By the beginning of the seventeenth century the situation in European mathematics was as follows. There were two algebras. The first was arithmetic, based on symbols created by the Europeans themselves and representing a substantial advance in comparison with the arithmetic of the ancients. The second algebra, geometric algebra, was part of geometry. It was taken, as was the whole of geometry, from the Greeks. The fundamentals were from Euclid's *Elements* and the further development came primarily from the works of Pappus of Alexandria and Apollonius, who had been thoroughly studied by that time. Nothing fundamentally new had been done in this field. We cannot say that there was no relationship at all between these two algebras; equations of degrees higher than the first could only receive geometric interpretation, for where else could squares, cubes, and higher degrees of an unknown number occur but in computing areas, volumes, and manipulations of line segments related by complex systems of proportions? The very names of the second and third degrees, the square and the cube, illustrate this very eloquently. Nonetheless, the gap between the concepts of quantity (or magnitude) and number remained and in full conformity with the Greek canon only geometric proofs were considered real. When geometric objects—lengths, areas, and volumes—appeared in equations they operated either as geometric quantities or as concrete numbers. Geometric quantities were thought of as necessarily something spatial and, because of incommensurability, not reducible to a number.

This was the situation that René Descartes (1596–1650), one of the greatest thinkers who has ever lived, encountered.

■ WHAT DID DESCARTES DO?

DESCARTES' ROLE as a philosopher is generally recognized. But when Descartes as a mathematician is discussed it is usually indicated that he "refined algebraic notations and created analytic geometry." Sometimes it is added that at approximately the same time the basic postulates of analytic geometry were proposed, independently of Descartes, by his countryman Pierre de Fermat (1601–1665), while Vieta had already made full use of algebraic symbols. It comes out, thus, that there is no special cause to praise Descartes the mathematician, and in fact many authors writing about the history of mathematics do not give him his due. However, Descartes carried out a revolution in mathematics. He created something incomparably greater than analytic geometry (understood as the theory of curves on a plane). What he created was a new approach to describing the phenomena of reality: the modern mathematical language.

It is sometimes said that Descartes "reduced geometry to algebra," which means, of course, numerical algebra, arithmetic algebra. This is a flagrant mistake. It is true that Descartes overcame the gap between quantity and number, between geometry and arithmetic. He did not achieve this by reducing one language to the other, however; he created a new language, the language of algebra. Not arithmetic algebra, not geometric algebra, simply algebra. In syntax the new language coincided with arithmetic algebra, but by semantics it coincided with geometric. In Descartes' language the symbols do not designate number or quantities, but relations of quantities. This is the essence of the revolution carried out by Descartes.

The modern reader will perhaps shrug his shoulders and think, "So what? Could this logical nuance really have been very important?" As it turns out, it was. It was precisely this "nuance" that had prevented the Greeks from taking the next step in their mathematics.

We have become so accustomed to placing irrational numbers together with rational ones that we are no longer aware of the profound difference which exists between them. We write $\sqrt{2}$ as we write $^4/_5$, and we call $\sqrt{2}$ a number and, when necessary, substitute an approximate value for it. And there is no way we can understand why the ancient Greeks responded with such pain to the incommensurability of line segments. But if we think a little, we cannot help agreeing with the Greeks that $\sqrt{2}$ is not a number. It can be represented as an infinite process which generates the sequential characters of expansion of $\sqrt{2}$ in decimal fractions. It can also be pictured in the form of a boundary line in the field of rational numbers—one that divides rational numbers into two classes: those which are less than $\sqrt{2}$ and those which are greater than $\sqrt{2}$. In this case the rule is very simple: the rational number a belongs to the first class if $a^2 < 2$ and to the second where this is not true. Finally, $\sqrt{2}$ can be pictured in the form of a relation between two line segments, between the diagonal of a square and its side in the particular case. These representations are equivalent to one another, but they are not at all equivalent to the representation of the whole or fractional number.

This by no means implies that we are making a mistake or not being sufficiently strict when we deal with $\sqrt{2}$ as a number. The goal of mathematics is to create linguistic models of reality, and all means which lead to this goal are good. Why shouldn't our language contain characters of the type $\sqrt{2}$ in addition to ones such as $^4/_5$? "It is my language and I will do what I want to with it." The only important thing is that we be able to interpret these characters and perform linguistic conversions on them. But we are able to interpret $\sqrt{2}$. In practical computations the first of the three representations in the preceding paragraph may serve as the basis of interpretation, while in geometry the third can be used. We can also carry out other computations with them. All that remains now is to refine the terminology. Let us stipulate that we shall use the term *rational* numbers for what were formerly called numbers, name the new objects *irrational* numbers, and use the term *numbers* for both (*real* numbers according to modern mathematical terminology). Thus, in the last analysis, there is no difference in principle between $\sqrt{2}$ and $^4/_5$ and we have proved wiser than the Greeks. This wisdom was brought in as contraband by all those who operated with the symbol $\sqrt{2}$ as a number, while recognizing that it was "irrational." It was Descartes who substantiated this wisdom and established it as law.

■ THE RELATION AS AN OBJECT

THE GREEKS' failure to create algebra is profoundly rooted in their philosophy. They did not even have arithmetic algebra. Arithmetic equations held little interest for them; after all, even quadratic equations do not, generally speaking, have exact numerical solutions. And approximate calculations and everything bound up with practical problems were uninteresting to them. On the other hand, the solution could have been found by geometric construction! But even if we assume that the Greek mathematicians of the Platonic school were familiar with arithmetic letter symbols it is difficult to imagine that they would have performed Descartes' scientific feat. To the Greeks, relations were not ideas and therefore did not have real existence. Who would ever think of using a letter to designate something that does not exist? The Platonic idea is a generalized image, a form, a characteristic; it can be pictured in the imagination as a more or less generalized object. All this is primary and has independent existence, an existence even more real than that of things perceived by the senses. But what is a relation of line segments? Try to picture it and you will immediately see that what you are picturing is precisely two line segments, not any kind of relation. The concept of the relation of quantities reflects the process of measuring one by means of the other. But the process is not an idea in the Platonic sense; it is something secondary that does not really exist. Ideas are eternal and invariable, and by this alone have nothing in common with processes.

Interestingly, the concept of the relation of quantities, which reflects characteristics of the measurement process, was introduced in strict mathematical form as early as Eudoxus and was included in the fifth book of Euclid's *Elements*. This was exactly the concept Descartes used. But the relation as an *object* is not found in Eudoxus or in later Greek mathematicians; after being introduced it slowly gave way to the proportion, which it is easy to picture as a characteristic of four line segments formed by two parallel lines intersecting the sides of an angle.

The concept of the relation of quantities is a linguistic construct, and quite a complex one. But Platonism did not permit the introduction of constructs in mathematics; it limited the basic concepts of mathematics to precisely representable static spatial images. Even fractions were considered somehow irregular by the Platonic school from the point of view of real mathematics. In *The Republic* we read: "If you want to divide a unit, learned mathematicians will laugh at you and will not permit you to do it; if you change a unit for small pieces of money they believe it has been turned into a set and are careful to avoid viewing the unit as consisting of parts rather than as a whole." With such an attitude toward rational numbers, why even talk about irrational ones!

We can briefly summarize the influence of Platonic idealism on Greek mathematics as follows. By recognizing mathematical statements as objects to work with, the Greeks made a metasystem transition of enormous importance, but then they immediately objectivized the basic elements of mathematical statements and began to view them as part of a nonlinguistic reality, the "world of ideas." In this way they closed off the path to further escalation of critical thinking, to becoming aware of the basic elements (concepts) of mathematics as phenomena of language and to creating increasingly more complex mathematical constructs. The development of mathematics in Europe was a continuous liberation from the fetters of Platonism.

■ DESCARTES AND FERMAT

IT IS VERY INSTRUCTIVE to compare the mathematical work of Descartes and Fermat. As a mathematician Fermat was as gifted as Descartes, perhaps even more so. This can be seen from his remarkable works on number theory. But he was an ardent disciple of the Greeks and continued their traditions. Fermat set forth his discoveries on number theory in remarks in the margins of Diophantus' *Arithmetica*. His works on geometry originated as the result of efforts to prove certain statements referred to by Pappus as belonging to Apollonius, but presented without proofs. Reflecting on these problems, Fermat began to systematically represent the position of a point on a plane by the lengths of two line segments: the *abscissa* and the *ordinate* and represent the curve as an equation relating these segments. This idea was not at all new from a geometric point of view; it was a pivotal idea not only in Apollonius, but even as far back as Archimedes, and it originates with even more ancient writers. Archimedes describes conic sections by their "symptoms," that is, the proportions which connect the abscissas and ordinates of the points. As an example, let us take an ellipsis with the longer (major) axis AB (see figure 11.3). Perpendicular line PQ, which is dropped from a certain point of the ellipsis P to axis AB, is called the ordinate, and segments AQ and QB are the abscissas of this point (both terms are Latin translations of Archimedes' Greek terms). The ratio of the area of a square constructed on the ordinate to the area of the rectangle constructed on the two abscissas is the same for all points P lying on the ellipsis. This is the "symptom" of the ellipsis, that is, in essence, its equation. It can be written as $Y^2 : X_1 X_2 = $ const. Analogous symptoms are established for the hyperbola and parabola. How is this not a system of coordinates?

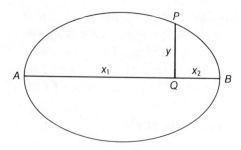

Figure 11.3.

Unlike the ancients, Fermat formulates the symptoms as equations in Vieta's language, not in the form of proportions described by words. This makes conversions easier; specifically, it can be seen immediately that it is more convenient to leave one abscissa than two. But the approach continues to be purely geometric, spatial.

Fermat set forth his ideas in the treatise "Ad locos planos et solidos isagoge" (Introduction to Plane and Solid Loci). This work was published posthumously in 1679, but it had been known to French mathematicians as early as the 1630s, somewhat earlier than Descartes' mathematical works.

Descartes' famous *Géométrie* came out in 1637. Descartes was not, of course, at all influenced by Fermat (it is unknown whether he even read Fermat's treatise); Descartes' method took shape in the 1620s, long before the *Géométrie* was published. Nonetheless, the properly geometrical ideas of Descartes and Fermat are practically identical. But Descartes created a new algebra based on the concept of the relation of geometric quantities. In Vieta only similar quantities can be added and subtracted and coefficients must include an indication of their geometric nature. For example, the equation which we would write as

$$A^3 + BA = D,$$

Vieta wrote as follows:

A cubus + B planum in A aequatur D solido.

This means, the cube with edge A added to area B multiplied by A is equal to the volume D. Vieta and Fermat are intellectual prisoners of the Greek geometric algebra. Descartes breaks with it decisively. The relations Descartes' algebra deals with are not geometric, spatial objects, but theoretical concepts, "numbers." Descartes is not restricted by the requirement for uniformity of things being added or the general requirement of a spatial interpretation; he understands raising to a power as repeated multiplication and indicates the number of factors by a small digit above and to the right. Descartes' symbolism virtually coincides with our modern system.

■ THE PATH TO DISCOVERY

FERMAT WAS ONLY a mathematician; Descartes was above all a philosopher. His reflections went far beyond mathematics and dealt with the problems of the essence of being and knowledge. Descartes was the founder of the philosophy of rationalism, which affirms the human being's unlimited ability to understand the world on the basis of a small number of intuitively clear truths and proceeding forward step by step using definite *rules* or *methods*. These two words are key words for all Descartes' philosophy. The name of his first philosophical composition was *Regulae ad directionem ingenii* (Rules for the Direction of the Mind), and his second was *Discours de la méthode* (Discourse on the Method). The *Discours de la méthode* was published in 1637 in a single volume with three physico-mathematical treatises: "la Dioptrique" (the Dioptric), "les Météores" (Meteors), and "la Gémétrie" (Geometry). The *Discours* preceded them as a presentation of the philosophical principles on which the following parts were based. In this *Discours* Descartes proposes the following four principles of investigation:

The first of these was to accept nothing as true which I did not clearly recognize to be so: that is to say, carefully to avoid precipitation and prejudice in judgments, and to accept in them nothing more than what was presented to my mind so clearly and distinctly that I could have no occasion to doubt it.

The second was to divide up each of the difficulties which I examined into as many parts as possible, and as seemed requisite in order that it might be resolved in the best manner possible.

The third was to carry on my reflections into order, commencing with objects that were the most simple and easy to understand, in order to rise little by little, or by degrees, to knowledge of the most complex, assuming an order, even if a fictitious one, among those which do not follow a natural sequence relatively to one another.

The last was in all cases to make enumerations so complete and reviews so general that I should be certain of having omitted nothing.[4]

Descartes arrived at his mathematical ideas guided by these principles. Here is how he himself describes his path in *Discours de la méthode:*

And I have not much trouble in discovering which objects it was necessary to begin with, for I already knew that it was with the most simple and those most easy to apprehend. Considering also that of all those who have hitherto sought for the truth in the Sciences, it has been the mathematicians alone who have been able to succeed in making any demonstrations, that is to say producing reasons which are evident and certain, I did not doubt that it had been by means of a similar kind that they carried on their investigations. . . . But for all that I had no intention of trying to master all those particular sciences that receive in common the name of Mathematics; but observing that, although their objects are different, they do not fail to agree in this, that they take nothing under consideration but the various relationships or proportions which are present in these objects, I thought that it would be better if I only examined those proportions in their general aspect, and without viewing them otherwise than in the objects which would serve most to facilitate a knowledge of them. Not that I should in any way restrict them to these objects, for I might later on all the more easily apply them to all other objects to which they were applicable. Then, having carefully noted that in order to comprehend the proportions I should sometimes require to consider each one in particular, and sometimes merely keep them in mind, or take them in groups, I thought that in order the better to consider them in detail, I should picture them in the form of lines, because I could find no method more simple nor more capable of being distinctly represented to my imagination and senses. I considered, however, that in order to keep them in my memory or to embrace several at once, it would be essential that I should explain

[4] *Descartes, Spinoza,* Great Books of the Western World, Encyclopedia Britanica Inc., Vol. 31, 1952, p. 47.

them by means of certain formulas, the shorter the better. And for this purpose it was requisite that I should borrow all that is best in Geometrical Analysis and Algebra, and correct the errors of the one by the other.[5]

We can see from this extremely interesting testimony that Descartes was clearly aware of the semantic novelty of his language based on the abstract concept of the relation and applicable to all the phenomena of reality. Lines serve only to illustrate the concept of the relation, just as a collection of little sticks serves to illustrate the concept of number. In their mathematical works Descartes and subsequent mathematicians have followed tradition and used the term "quantity" for that which is designated by letters, but semantically these are not the spatial geometric quantities of the Greeks but rather their relations. In Descartes the concept of quantity is just as abstract as the concept of number. But of course, it cannot be reduced to the concept of number in the exact meaning of the word, that is, the rational number. Explaining his notations in the *Géométrie,* Descartes points out that they are similar (but not identical) to the notations of arithmetic algebra:

> Just as arithmetic consists of only four or five operations, namely, addition, subtraction, multiplication, division, and the extraction of roots, which may be considered a kind of division, so in geometry, to find required lines it is merely necessary to add or subtract other lines; or else, taking one line which I shall call unity in order to relate it as closely as possible to numbers, and which can in general be chosen arbitrarily, and having given two other lines, to find a fourth line which shall be to one of the given lines as the other is to unity (which is the same as multiplication); or, again, to find a fourth line which is to one of the given lines as unity is to the other (which is equivalent to division); or, finally, to find one, two, or several mean proportionals between unity and some other line (which is the same as extracting the square root, cube root, etc., of the given line). And I shall not hesitate to introduce these arithmetical terms into geometry, for the sake of greater clearness.[6]

[5] Descartes, *ibid.,* p. 47.   [6] Descartes, *ibid.,* p. 295.

The semantics of Descartes' algebraic language are much more complex than the semantics of the arithmetic and geometric languages which rely on graphic images. The use of such a language changes one's view of the relation between language and reality. It is discovered that the letters of mathematical language may signify not only numbers and figures, but also something much more abstract (to be more precise, "constructed"). This is where the invention of new mathematical languages and dialects and the introduction of new constructs began. The precedent was set by Descartes. Descartes in fact laid the foundation for describing the phenomena of reality by means of formalized symbolic languages.

The immediate importance of Descartes' reform was that it untied the hands of mathematicians to create, in abstract symbolic form, the infinitesimal analysis, whose basic ideas in geometric form were already known to the ancients. If we go just half a century from the publication date of the *Géométrie* we find ourselves in the age of Leibnitz and Newton, and 50 more years brings us to the age of Euler.

The history of science shows that the greatest glory usually does not go to those who lay the foundations and, of course, not to those who work on the small finishing touches; rather it goes to those who are the first in a new line of thought to obtain major results which strike the imagination of their contemporaries or immediate descendants. In European physico-mathematical science this role was played by Newton. But as Newton said, "If I have seen further than Descartes, it is by standing on the shoulders of giants." This is, of course, evidence of the modesty of a brilliant scientist, but it is also a recognition of the debt of the first great successes to the pioneers who showed the way. The apple which made Newton famous grew on a tree planted by Descartes.

CHAPTER TWELVE
From Descartes to Bourbaki

■ FORMALIZED LANGUAGE

"THE NEXT STOP IS APRELEVKA STATION," a hoarse voice announces through the loudspeaker. "I repeat, Aprelevka Station. The train does not have a stop at Pobeda Station."

You are riding a commuter train on the Kiev Railroad, and because you have forgotten to bring a book and there is nothing for you to do, you begin reflecting on how carelessly we still treat our native language. Really, what an absurd expression, "does not have a stop." Wouldn't it be simpler to say "does not stop"? Oof, these bureaucratic, governmental expressions! People write about it all the time, but it hasn't done any good yet.

If you do not get off at Aprelevka, however, and you have time for further reflection you will see that this is by no means a matter of a careless attitude toward our native language; in fact, "does not have a stop" does not mean quite the same thing as "does not stop." The concept of the stop in railroad talk is not the same as the concept of ceasing movement. The following definition, not too elegant but accurate enough, can be given: a stop is a deliberate cessation of the train's movement accompanied by the activities necessary to ensure that passengers get on and off the train. This is a very important concept for railroad workers and it is linked to the noun "stop," not to the verb "to stop." Thus, if the engineer stopped the train but did not open the pneumatic doors, the train "stopped" but it did not "have a stop."

The railroad worker who made the announcement did not, of course, perform such a linguistic analysis. He simply used the ordinary professional term, which enabled him to express his thought exactly, even if it seemed somewhat clumsy to a nonprofessional. This is an instance of a very common phenomenon: when language is used for comparatively narrow professional purposes there is a tendency to limit the number of terms used and to give them more precise and constant meanings. We say the language is *formalized*. If this process is carried through to its logical conclusion the language will be completely formalized.

The concept of a formalized language can be defined as follows. Let us refer to our diagram of the use of linguistic models of reality (see figure 9.5) and put the question: how is the conversion $L_1 \rightarrow L_2$ performed, on what information does it depend? We can picture two possibilities:

1. The conversion $L_1 \rightarrow L_2$ is determined exclusively by linguistic objects L_i which participate in it and do not depend on those nonlinguistic representations S_i which correspond to them according to the semantics of the language. In other words, the linguistic activity depends only on the "form" of the language objects not on their "content" (meaning).

2. The result of the conversion of linguistic object L_i depends less on the type of object L_i itself than on representation S_i it generates in the person's mind, on the associations in which it is included, and therefore on the person's personal experience of life.

In the first case we call the language *formalized*, while in the second case it is *unformalized*. We should emphasize that complete formalization of a language does not necessarily mean complete algorithmization of it, the situation where all linguistic activity amounts to fulfilling precise and unambiguous prescriptions as a result of which each linguistic object L_1 is converted into a completely definite object L_2. The rules of conversion $L_1 \rightarrow L_2$ can be formalized as more or less rigid constraints and leave a certain freedom of action; the only important thing is that these constraints depend on the type of object L_1 and potential objects L_2 by themselves alone and not on the meanings of the linguistic objects.

The definition we have given of a formalized language applies to the case where language is used to create models of reality. When a language serves as a means of conveying control information (the language of orders) there is a completely analogous division into two possible types of responses:

1. The person responds in a strictly formal manner to the order, that is, his actions depend only on the information contained in the text of the order, which is viewed as an isolated material system.

2. The person's actions depend on those representations and associations the order evokes in him. Thus, he actually uses much more information than that contained in the text of the order.

There is no difference in principle between the language of orders and the language of models. The order "Hide!" can be interpreted as the model "If you don't hide your life is in danger." The difference between the order and the model is a matter of details of information use. In both cases the formalized character of the language leads to a definite division of syntax and semantics, a split between the material linguistic objects and the representations related to them; the linguistic objects acquire the characteristics of an independent system.

Depending on the type of language which is used we may speak of informal and formal thinking. In informal thinking, linguistic objects are primarily important to the extent that they evoke definite sets of representations in us. The words here are strings by which we extract from our memory particles of our experience of life; we relive them, compare them, sort through them, and so on. The result of this internal work is the conversion of representations $S_1 \rightarrow S_2$, which models the changes $R_1 \rightarrow R_2$ in the environment. But this does not mean that informal thinking is identical to nonlinguistic thinking. In the first place, by itself the dismembering of the stream of perceptions depends on a system of concepts fixed in language. In the second place, in the process of the conversion $S_1 \rightarrow S_2$ the "natural form" of the linguistic object, the word, plays a considerable part. Very often we use associations among words, not among representations. Therefore the formula for nonformal thinking can be represented as follows: $(S_1, L_1) \rightarrow (S_2, L_2)$,

In formal thinking we operate with linguistic objects as if they were certain independent and self-sufficient essences, temporarily forgetting their meanings and recalling them only when it is necessary to interpret the result received or refine the initial premises. The formula for formal thinking is as follows: $S_1 \rightarrow L_1 \rightarrow L_2 \rightarrow S_2$.

In order for formal thinking to yield correct results, the semantic system of the language must possess certain characteristics we describe by such terms as "precision," "definiteness," and "lack of ambiguity." If the semantic system does not possess these characteristics, we shall not be able to introduce such formal conversions $L_1 \rightarrow L_2$ in order that, by using them, we may always receive a correct answer. Of course, it is possible to establish the formal rules of conversions somehow and thus obtain a formalized language, but this will be a language that sometimes leads to false conclusions. Here is an example of a deduction which leads to a false result because of ambiguity in the semantic system:

Vanya is a gypsy.
The gypsies came to Europe from India.
Therefore, Vanya came to Europe from India.

In practice, thus, semantic precision and syntactical formalization are inseparable, and a language that satisfies both criteria is called formalized. But the leading criterion is the syntactical one, for the very concept of a precise semantic system can be defined strictly only through syntax. And indeed, the semantic system is precise if it is possible to establish formalized syntax which yields only true models of reality.

■ THE LANGUAGE MACHINE

BECAUSE the syntactical conversions $L_1 \rightarrow L_2$ within the framework of a formalized language are determined entirely by the physical type of objects L_i, the formalized language is in essence a machine that produces different changes of symbols. For a completely algorithmized language, such as the language of arithmetic, this thesis is perfectly obvious and is illustrated by the existence of machines in the ordinary, narrow sense of the word (calculators and electronic computers) that carry out arithmetic algorithms. If the rules of conversion are constraints only, it is possible to construct an algorithm that determines whether the conversion $L_1 \rightarrow L_2$ is proper for given L_1 and L_2. It is also possible to construct an algorithm (a "stupid" one) which for a given L_1 begins to issue all proper results for L_2 and continues this process to infinity if the number of possible L_2 is unlimited. In both cases we are dealing with a certain *language machine*, that can work without human intervention.

The formalization of a language has two direct consequences. In the first place, the process of using linguistic models is simplified because precise rules for converting $L_1 \rightarrow L_2$ appear. In the extreme case of complete algorithmization, this conversion can generally be carried out automatically. In the second place, the linguistic model becomes independent of the human brain which created it, and becomes an objective model of reality. Its semantic system reflects, of course, concepts that have emerged in the process of the development of the culture of human society, but in terms of syntax it is a language machine that could continue to work and preserve its value as a model of reality even if the entire human race were to suddenly disappear. By studying this model an intelligent being with a certain knowledge of the object of modeling would probably be able to reproduce the semantic system of the language by comparing the model to his own knowledge. Let us suppose that people have built a mechanical model of the Solar System in which the planets are represented by spheres of appropriate diameters revolving on pivots around a central sphere, representing the sun, in appropriate orbits with appropriate periods. Then let us suppose that this model has fallen into the hands (perhaps the tentacles?) of the inhabitants of a neighboring stellar system, who know some things about our Solar System—for example, the distances of some planets from the sun or the times of their revolutions. They will be able to understand what they have in front of them, and they will receive additional information on the Solar System. The same thing is true of scientific theories, which are models of reality in its different aspects, built with the material of formalized symbolic language. Like a mechanical model of the Solar System, each scientific theory can in principle be deciphered and used by any intelligent beings.

■ FOUR TYPES OF LINGUISTIC ACTIVITY

Language can be characterized not only by the degree of its formalization but also by the degree of its abstraction, which is measured by the abundance and complexity of the linguistic constructs it uses. As we noted in chapter 7, it would be more correct to speak of the "construct quality" of a language rather than of its abstractness, but the former term [the Russian "konstruktnost' "] has not yet been accepted. Therefore we shall use the term "abstractness." We shall call a language which does not use constructs or uses only those of the very lowest level "concrete," and we shall call a language which does use complex constructs "abstract." Although this is a conditional and relative distinction, its meaning is nonetheless perfectly clear. And it does not depend on dividing languages into formalized and unformalized, which are different aspects of language. By combining all these aspects we obtain four types of languages used in the four most important spheres of linguistic activity. They can be arranged according to the table below:

	Concrete Language	Abstract Language
Unformalized Language	Art	Philosophy
Formalized Language	Descriptive sciences	Theoretical sciences (mathematics)

Neither the vertical nor the horizontal division is strict and unambiguous; the differences are more of a quantitative nature. There are transitional types on the boundaries between these "pure" types of language.

Art is characterized by unformalized and concrete language. Words are important only as symbols which evoke definite complexes of representations and emotions. The emotional aspect is ordinarily decisive, but the cognitive aspect is also very fundamental. In the most significant works of art these aspects are inseparable. The principal expressive means is the image, which may be synthetic but always remains concrete.

Moving leftward across the table, we come next to philosophy, which is characterized by abstract, informal thinking. The combination of an extremely high degree of constructs among the concepts used and an insignificant degree of formalization requires great effort by the intuition and makes philosophical language unquestionably the most difficult of the four types of language. When art raises abstract ideas it comes close to philosophy. On the other hand, philosophy will use the artistic image now and again to stimulate the intuition, and here it borders on art.

On the bottom right half of our table we find the theoretical sciences, characterized by an abstract and formalized language. Science in general is characterized by formalized language; the difference between the descriptive and theoretical sciences lies in a different degree of use of concept-constructs. The language of descriptive science must be concrete and precise; formalization of syntax by itself does not play a large part, but rather acts as a criterion of the precision of the semantic system (logical consistency of definitions, completeness of classifications, and so on).

The models of the world given by the descriptive sciences [bottom left of the table] are expressed in terms of ordinary neuronal concepts or concepts with a low degree of construct usage and, properly speaking, as models they are banal and monotypic: if some particular thing is done (for example, a trip to Australia or cutting open the abdominal cavity of a frog) it will be possible to see some other particular thing. On the other hand, the whole essence of the theoretical sciences is that they give fundamentally new models of reality: scientific theories based on concept-constructs not present at the neuronal levels. Here the formalization of syntax plays the decisive part. The most extreme of the theoretical sciences is mathematics, which contains the most complex constructs and uses a completely formalized language. Properly speaking mathematics is the formalized language used by the theoretical sciences.

Moving back up from the descriptive sciences we are again in the sphere of art. Somewhere on the border between the descriptive sciences and art lies the activity of the journalist or naturalist-writer.

■ SCIENCE AND PHILOSOPHY

ALTHOUGH THE LANGUAGE of science is formalized, scientists cannot restrict themselves to purely formal thinking. The use of a complete and finished theory does indeed demand formal operations that do not go outside the framework of a definite language, but the creation of a new theory always involves going beyond the formal system; it is always a metasystem transition of greater or lesser degree.

Of course, we certainly cannot say that everyone who does not break down old formalisms is working on banal and uncreative things. This applies only to those who operate in accordance with already available algorithms, essentially performing the functions of a language machine. But fairly complex formal systems cannot be algorithmized and they offer a broad area for creative activity. Actions within the framework of such a system can be compared to playing chess. In order to play chess well one must study for a long time, memorize different variations and combinations, and acquire a specific chess intuition. In the same way the scientist who is dealing with a complex formalized language (that is to say, with mathematics either pure or applied) develops in himself, through long study and training, an intuition for his language, often a very narrow one, and obtains new theoretical results. This is, of course, activity which is both noble and creative.

All the same, going beyond the old formalism is an even more serious creative step. If the scientists we were discussing above could be called chess-player–scientists, then the scientists who create new formalized languages and theories can be called philosopher-scientists. We saw an example of these two types of scientist in our discussion of Fermat and Descartes in chapter 11. The concepts of new theories do not emerge in precise and formalized form from a vacuum. They become crystallized gradually, during a process of abstract but not formalized thinking—i.e., philosophical thinking. And whereas here too intuition is required, it is of a different type—

philosophical. "The sciences," Descartes wrote in his *Discours de la méthode* "borrow their principles from philosophy."

The creation of fundamental scientific theories lies in the borderline area between philosophy and science. As long as a scientist operates with conventional concepts within the framework of conventional formalized language he does not need philosophy. He is like the chess player who pictures the same pieces on the same board, but solves different problems. And he does obtain new results, relying on his intuition for chess. But in this he will never, in his game of chess, go beyond the limits inherent in his language. To improve language itself, to formalize what has not yet been formalized means to go into philosophy. If a new theory does not contain this element it is only a consequence of old theories. It can be said that the amount of what is new in any theory corresponds exactly to the amount of philosophy in it.

From the above discussion the importance of philosophy for the activity of the scientist is clear. In the *Dialectic of Nature*, F. Engels wrote:

> Naturalists imagine that they are free of philosophy when they ignore or downgrade it. But because they cannot take a step without thinking, and thinking demands logical categories and they borrow these categories uncritically either from the everyday, general consciousness of so-called educated people among whom the remnants of long-dead philosophical systems reign, from crumbs picked up in required university courses in philosophy (which are not only fragmentary views, but also a hodgepodge of the views of people affiliated with the most diverse and usually the most despicable schools), or from uncritical and unsystematic reading of every kind of philosophical works—in the end they are still subordinate to philosophy but, unfortunately, it is usually the most despicable philosophy and those who curse philosophy most of all are slaves to the worst vulgarized remnants of the worst philosophical systems.[1]

That sounds amazingly modern!

[1] Engels, F. *Dialektika prirody* (The Dialectic of Nature), Gospolitizdat Publishing House, 1955, p. 165.

■ FORMALIZATION AND THE
METASYSTEM TRANSITION

THE CONVERSION of language, occurring as a result of formalization, into a reality independent of the human mind which creates it has far-reaching consequences. The just-created language machine (theory), as a part of the human environment, becomes an object of study and description by means of the new language. In this way a metasystem transition takes place. In relation to the described language the new language is a metalanguage and the theories formulated in this language and concerned with theories in the language-object are meta-theories. If the metalanguage is formalized, it may in turn become an object of study by means of the language of the next level and this metasystem transition can be repeated without restriction.

In this way, the formalization of a language gives rise to the stairway effect (see chapter 5). Just as mastering the general principles of making tools to influence objects gives rise to multiple repetitions of the metasystem transition and the creation of the hierarchical system of industrial production, so mastering the general principle of describing (modeling) reality by means of a formalized language gives rise to creation of the hierarchical system of formalized languages on which the modern exact sciences are based. Both hierarchies have great height. It is impossible to build a jet airplane with bare hands. The same thing is true of the tools needed to build an airplane. One must begin with the simplest implements and go through the whole hierarchy of complexity of instruments before reaching the airplane. In exactly the same way, in order to teach the savage quantum mechanics, one must begin with arithmetic.

■ THE LEITMOTIF OF THE NEW MATHEMATICS

THE ESSENCE of what occurred in mathematics in the seventeenth century was that the general principle of using formalized language was mastered. This marked the beginning of movement up the stairway; it led to grandiose achievements and continues to the present day. It is true that this principle was not formulated so clearly then as now, and the term "formalized language" did not appear until the twentieth century. But such a language was in face used. As we saw, Descartes' reform was the first step along this path. The works of Descartes, in particular the quotations given above, show that this step was far from accidental; rather it followed from his method of learning the laws of nature which, if we put it in modern terms, is the method of creating models using formalized language. Descartes was aware of the universality of his method and its mathematical character. In the *Regulae ad directionem ingenii* he expresses his confidence that there must be "some general science which explains everything related to order and measure without going into investigation of any particular objects." This science, he writes, should be called "universal mathematics."

Another great mathematician-philosopher of the seventeenth century, G. Leibnitz (1646–1716), understood fully the importance of the formalization of language and thinking. Throughout his life Leibnitz worked to develop a symbolic calculus to which he gave the Latin name *characteristica universalis*. Its goal was to express all clear human thoughts and reduce logical deduction to purely mechanical operations. In one of his early works Leibnitz states, "The true method should be our Ariadne's thread, that is, a certain palpable and rough means which would guide the reason like lines in geometry and the forms of operations prescribed for students of arithmetic. Without this our reason could not make the long journey without getting off the road." This essentially points out the role of formalized language as the material fixer of concept-constructs—i.e., its main role. In his historical essay on the foundations of mathematics [2] N. Bourbaki writes:

> The many places in the works of Leibnitz where he mentions his gran-
> diose project and the progress which would follow upon its realization
> show how clearly he understood formalized language as a pure combi-
> nation of characters in which only their coupling is important, so that a
> machine will be able to derive all theorems and it will be possible to

[2] Bourbaki, N. *Elements d'histoire des mathematiques*. Paris: Hermann. The quote is from the first essay, in the section "Formalization of Logic."

resolve all incomplete or mistaken understanding by simple calcula-tion. Although such hopes might seem excessive, it must be admitted that it was in fact under the constant influence of them that Leibnitz created a significant share of his mathematical writings, above all his works on the symbolism of infinitesimal calculus. He himself was very well aware of this and openly linked his ideas of introducing indexes and determinants and his draft of the ''geometric calculus'' to his ''charateristica.'' But he felt that his most significant work would be symbolic logic. . . . And although he was not able to create such a calculus, at least he started work to carry out his intention three times.

Leibnitz's ideas on the *characteristica universalis* were not elab-orated in his day. The work of formalizing logic did not get un-derway until the second half of the nineteenth century. But Leibnitz's ideas are testimony to the fact that the principle of describing reality by means of formalized logic is an inborn characteristic of European mathematics, and has always been the source of its development, even though different authors have been aware of this to different degrees.

It is not our purpose to set forth the history of modern mathe-matics or to give a detailed description of the concepts on which it is based; a separate book would be required for that. We shall have to be satisfied with a brief sketch that only touches that aspect of mathe-matics which is most interesting to us in this book—specifically, the system aspect.

The leitmotif in the development of mathematics during the last three centuries has been the gradually deepening awareness of mathe-matics as a formalized language and the resulting growth of multiple levels in it, occurring through metasystem transitions of varying scale.

We shall now review the most important manifestations of this process; they can be called variations on a basic theme, performed on different instruments and with different accompaniment. Simulta-neously with upward growth in the edifice of mathematics there was an expansion of all its levels, including the lowest one—the level of applications.

■ "NONEXISTENT" OBJECTS

WE HAVE ALREADY spoken of "impossible" numbers—irrational, negative, and imaginary numbers. From the point of view of Platonism the use of such numbers is absolutely inadmissible and the corresponding symbols are meaningless. But Indian and Arabic mathematicians began to use them in a minor way, and then in European mathematics they finally and irreversibly took root and received reinforcement in the form of new "nonexistent" objects, such as an infinitely remote point of a plane. This did not happen all at once, though. For a long time the possibility of obtaining correct results by working with "nonexistent" objects seemed amazing and mysterious. In 1612 the mathematician Clavius, discussing the rule that "a minus times a minus yields a plus" wrote: "Here is manifested the weakness of human reason which is unable to understand how this can be true." In 1674, discussing a certain relation between complex numbers, Huygens remarked: "There is something incomprehensible to us concealed here." A favorite expression of the early eighteenth century was the "incomprehensible riddles of mathematics." Even Cauchy in 1821 had very dim notions of operations on complex quantities.[3]

The last doubts and uncertainties related to uninterpreted objects were cleared up only with the introduction of the axiomatic approach to mathematical theories and final awareness of the "linguistic nature" of mathematics. We now feel that there is no more reason to be surprised at or opposed to the presence of such objects in mathematics than to be surprised at or opposed to the presence of parts in a car in addition to the four wheels, which are in direct contact with the ground and set the car in motion. Complex numbers and objects like them are the internal "wheels" of mathematical models; they are connected with other "wheels," but not directly with the "ground," that is, the elements of nonliguistic reality. Therefore one may go right on and operate with them as *formal objects* (that is, characters

---

[3] This opinion and the quotations given above were taken from H. Weyl's book *The Philosophy of Mathematics* (Russian edition *O filosofii matematiki*, Moscow–Leningrad, 1934.)

written on paper) in accordance with their properties as defined by axioms. And there is no reason to grieve because you cannot go to the pastry shop and buy $\sqrt{-15}$ rolls.

■ THE HIERARCHY OF THEORIES

AWARENESS OF THE PRINCIPLE of describing reality by means of formalized language gives rise, as we have seen, to the stairway effect. Here is an example of a stairway consisting of three steps. Arithmetic is a theory we apply directly to such objects of nonlinguistic reality as apples, sheep, rubles, and kilograms of goods. In relation to it school algebra is a metatheory that knows only one reality—numbers and numerical equalities—while its letter language is a metalanguage in relation to the language of the numerals of arithmetic. Modern axiomatic algebra is a metatheory in relation to school algebra. It deals with certain objects (whose nature is not specified) and certain operations on these objects (the nature of the operations is also not specified). All conclusions are drawn from the characteristics of the operations. In the applications of axiomatic algebra to problems formulated in the language of school algebra, objects are interpreted as variables and operations are arithmetic operations. But modern algebra is applied with equal success to other branches of mathematics, for example to analysis and geometry.

A thorough study of mathematical theory generates new mathematical theories which consider the initial theory in its different aspects. Therefore, each of these theories is in a certain sense simpler than the initial theory, just as the initial theory is simpler than reality, which it always considers in some certain aspect. The models are dismembered and a set of simpler models is isolated from the complex one. Formally speaking, new theories are just as universal as the initial theory: they can be applied to any objects, regardless of their nature, if they satisfy the axioms. With the axiomatic approach different mathematical theories form what is, strictly speaking, a hierarchy of complexity, not of control. When we consider the models that in fact express laws of nature (the ones used in applications of mathematics), however, we see that mathematical theories are very clearly divided into levels according to the nature of the objects to which they are actually applied. Arithmetic and elementary geometry are in direct contact with nonlinguistic reality, but a certain theory of groups is used to create new physical theories from which results expressed in the language of algebra and analysis are extracted and then "put in numbers"; only after this are they matched with experimental results. This distribution of theories by levels corresponds overall to the order in which they arose historically, because they arose through successive metasystem transitions. The situation here is essentially the same as in the hierarchy of implements of production. It is possible to dig up the ground with a screwdriver, but that tool was not invented for this purpose and really is needed only by someone working with screws and bolts. Group theory can be illustrated by simple examples from everyday life or elementary mathematics, but it is really used only by mathematicians and theoretical physicists. A clerk in a store or an engineer in the field has no more use for group theory than the primitive has for a screwdriver.

■ THE AXIOMATIC METHOD

ACCORDING TO THE ancient Greeks, the objects of mathematics had real existence in the "world of ideas." Some of the properties of these objects seemed in the mind to be absolutely indisputable; they were declared axioms. Others, which were not so obvious, had to be proved using the axioms. With such an approach there was no great need to precisely formulate and to completely list all the axioms: if some "indisputable" attribute of objects is used in a proof, it is not that important to know whether it has been included in a list of axioms or not; the truth of the property being proved does not suffer. Although Euclid did give a list of definitions and axioms (including postulates) in his *Elements,* as we saw in chapter 10, now and again he used assumptions which are completely obvious intuitively but not included in the list of axioms. As for his definitions, there are more of them than there are objects defined, and they are completely unsuitable for use in the proof process. The list of definitions in the first book of the *Elements* begins as follows:

1. The point is that which does not have parts.
2. The line is a length without width.
3. The ends of lines are points.
4. A straight line is a line which lies the same relative to all its points.

There are a total of 34 definitions. The Swiss geometer G. Lambert (1728–1777) noted in this regard: "What Euclid offers in this abundance of definitions is something like a nomenclature. He really proceeds like, for example, a watchmaker or other artisan who is beginning to familiarize his apprentices with the names of the tools of his trade."

The trend toward formalization of mathematics generated a trend toward refinement of definitions and axioms. Leibnitz called attention to the fact that Euclid's construction of an equilateral triangle relies on an assumption that does not follow from the definitions and axioms (we reviewed this construction in chapter 10). But it was only the creation of non-Euclidean geometry by N. I. Lobachevsky (1792–1856), J. Bolyai (1802–1860), and K. Gauss (1777–1855) which brought universal recognition of the axiomatic approach to mathematical theories as the fundamental method of mathematics. At first Lobachevsky's "imaginary" (conceptual) geometry, like all "imaginary" phenomena in mathematics, encountered distrust and hostility. Soon the irrefutable fact of the existence of this geometry began to change the point of view of mathematicians concerning the relation between mathematical theory and reality. The mathematician could not refuse Lobachevsky's geometry the right to exist, because this geometry was proved to be noncontradictory. It is true that Lobachevsky's geometry contradicted our geometrical intuition, but with a sufficiently small parameter of spatial curvature it was indistinguishable from Euclidean geometry in small spatial volumes. As for the cosmic scale, it is not at all obvious that we can trust our intuition there, because our intuition forms under the influence of experience limited to small volumes. Thus we face two competing geometries and the guestion arises: which of them is "true"?

When we ponder this question it become clear that the word "true" is not placed in quotation marks without reason. Strictly speaking, the experiment cannot answer the question of the truth or falsehood of geometry; it can only answer the question of its usefulness or lack of usefulness, or more precisely its degree of usefulness, for there are perhaps no theories which are completely useless. The experiment deals with physical, not geometric, concepts. When we turn to the experiment we are forced to give some kind of interpretation to geometric objects, for example to consider that straight lines are realized by light beams. If we discover that the sum of the angles of a triangle formed by light beams is less than 180 degrees, this in no way means that Euclidean geometry is "false." Possibly it is "true," but the light is propagated not along straight lines but along arcs of circumference or some other curved lines. To speak more precisely, this experiment will demonstrate that light beams cannot be considered as Euclidean straight lines. Euclidean geometry itself will not be refuted by this. The same thing applies, of course, to non-Euclidean geometry also. The experiment can answer the question of whether the light beam is an embodiment of the Euclidean straight line or the Lobachevsky straight line, and this of course is an important argument in choosing one geometry or the other as the basis for physical theories. But it does not take away the right to existence of the geometry which "loses out." It may perhaps do better next time and prove very convenient for describing some other aspect of reality.

Such considerations led to a reevaluation of the relative importance of the nature of mathematical objects and their properties (including relations as properties of pairs, groups of three, and other such objects). Whereas formerly objects seemed to have independent, real existence while their properties appeared to be something secondary and derived from their nature, now it was the properties of the objects, fixed in axioms, which became the basis by which to define the specific nature of the given mathematical theory while the objects lost all specific characteristics and, in general, lost their "nature," which is to say, the intuitive representations necessarily bound up with them. In axiomatic theory the object is something which satisfies the axioms. The axiomatic approach finally took root at the turn of the twentieth century. Of course, intuition continued to be important as the basic (and perhaps only) tool of mathematical creativity, but it came to be considered that the final result of creative work was the completely formalized axiomatic theory which could be interpreted to apply to other mathematical theories or to nonlinguistic reality.

■ METAMATHEMATICS

THE FORMALIZATION of logic was begun (if we do not count Leibnitz's first attempts) in the mid-nineteenth century in the works of G. Boole (1815–1864) and was completed by the beginning of the twentieth century, primarily thanks to the work of Schroeder, C. S. Pierce, Frege, and Peano. The fundamental work of Russell and Whitehead, the *Principia Mathematica*, which came out in 1910, uses a formalized language which, disregarding insignificant variations, is still the generally accepted one today. We described this language in chapter 6, and now we shall give a short outline of the formalization of logical deduction.

There are several formal systems of logical deduction which are equivalent to one another. We shall discuss the most compact one. It uses just one logical connective, implication $\supset$, and one quantifier, the universal quantifier $\forall$. But then it includes a logical constant which is represented by the symbol 0 and denotes an identically false statement. Using this constant it is possible to write the negation of statement p as $p \supset 0$, and from negation and implication it is easy to construct the other logical connectives. The quantifier of existence is expressed through negation and the quantifier of generality, so our compressed language is equivalent to the full language considered in chapter 6.

The formal system (language machine) contains five axioms and two rules of inference. The axioms are the following:

A1. $p \supset (q \supset p)$
A2. $[p \supset (q \supset r)] \supset [(p \supset q) \supset (p \supset r)]$
A3. $[(p \supset 0) \supset 0] \supset p$
A4. $(\forall x)[p \supset q(x)] \supset [p \supset (\forall x)q(x)]$
A5. $(\forall x)q(x) \supset q(t)$

In this p, q, and r are any propositions; in **A4** and **A5** the entry $q(x)$ means that one of the free variables on which proposition q depends has been isolated; the entry $q(t)$ means that some term t has been substituted for this variable; finally, in **A4** it is assumed that variable x does not enter p as a free variable.

It is easy to ascertain that these axioms correspond to our intuition. Axioms **A1–A3** involve only propositional calculus and their truth can be tested by the truth tables of logical connectives. It turns out that they are always true, regardless of the truth values assumed by propositions p, q, and r. **A4** says that if $q(x)$ follows for any x from proposition p which does not depend on x, the truth of $q(x)$ for any x follows from p. **A5** is in fact a definition of the universal quantifier: if $q(x)$ is true for all x, then it is also true for any t.

The rules of inference may be written concisely in the following way:

$$MP\frac{p/p \supset q}{q}, GN\frac{p(x)}{(\forall \xi)p(\xi)}.$$

In this notation the premises are above the line and the conclusion is below. The first rule (which traditionally bears the Latin name *modus ponens*) says that if there are two premises, proposition p and a proposition which affirms that q follows from p, then we deduce proposition p as the conclusion. The second rule, the rule of generalization, is based on the idea that if it has been possible to prove a certain proposition $p(x)$, which contains free variable x, it may be concluded that the proposition will be true for any value of this variable.

The finite sequence of formulas $D = (d_1, d_2, \ldots, d_n)$ such that d_n coincides with q and each formula d_i is either a formula from a set of premises X, a logical axiom, or a conclusion obtained according to the rules of inference from the preceding formulas d_j is called the logical deduction of formula q from the set of formulas (premises) X. When we consider axiomatic theory, the aggregate of all axioms of the given theory figures as the set X and the logical deduction of a certain formula is its *proof*.

Thus, the formula's proof itself became a formal object, a definite type of formula (sequence of logical statements) and as a result the possibility of purely syntactical investigation of proofs as characteristics of a certain language machine emerges. This possibility was pointed out by the greatest mathematician of the twentieth century, David Hilbert (1862–1943), who with his students laid the foundations of the new school. Hilbert introduced the concept of the *metalanguage* and called the new school *metamathematics*. The term *metasystem* which we introduced at the start of the book (and which is now generally accepted) arose as a result of generalizing Hilbert's terminology. Indeed, the transition to investigating mathematical proofs by mathematical means is a brilliant example of a large-scale metasystem transition.

The basic goal pursued by the program outlined by Hilbert was to prove that different systems of axioms were consistent (noncontradictory). A system of axioms is called contradictory if it is possible to deduce from it a certain formula q and its negation $-q$. It is easy to show that if there is at least one such formula, that is to say if the theory is contradictory, then any formula can be deduced from it. For an axiomatic theory, therefore, the question of the consistency of the system of axioms on which it is based is extremely important. This question admits a purely syntactical treatment: is it possible from the given formulas (strings of characters), following the given formal rules, to obtain a given formal result? This is the formulation of the question from which Hilbert began; it then turned out that there are also other important characteristics of theories which can be investigated by syntactical methods. Many very interesting and important results, primarily of a negative nature, were obtained in this way.

■ THE FORMALIZATION OF SET THEORY

THE CONCEPT of the aggregate or set is one of the most fundamental concepts given to us by nature; it precedes the concept of number. In its primary form it is not differentiated into the concepts of the finite and infinite sets, but this differentiation appears very early; in any case, in very ancient written documents we can already find the concept of infinity and the infinite set. This concept was used in mathematics from ancient times on, remaining purely intuitive, taken as self-explanatory and not subject to special consideration, until Georg Cantor (1845–1918) developed his theory of sets in the 1870s. It soon became the basis of all mathematics. In Cantor the concept of the set (finite or infinite) continues to be intuitive. He defines it as follows: "By a set we mean the joining into a single whole of objects which are clearly distinguishable by our intuition or thought." Of course, this "definition" is no more mathematical than Euclid's "definition" that "The point is that which does not have parts." But despite such imprecise starting points, Cantor (once again, like the Greek geometers) created a harmonious and logically consistent theory with which he was able to put the basic concepts and proofs of mathematical analysis into remarkable order. ("It is simply amazing," writes Bourbaki, "what clarity is gradually acquired in his writing by concepts which, it seemed, were hopelessly confused in the classical conception of the 'continuum.' ") [4] In set theory mathematicians received a uniform method of creating new concept-constructs and obtaining proofs of their properties. For example, the real number is the set of all sequences of rational numbers which have a common limit; the line segment is a set of real numbers; the function is the set of pairs (x, f) where x and f are real numbers.

By the end of the nineteenth century Cantor's set theory had become recognized and was naturally combined with the axiomatic method. But then the famous "crisis of the foundations" of mathematics burst forth and continued for three decades. "Paradoxes," which is to say constructions leading to contradiction, were found in set theory. The first paradox was discovered by Burali-Forti in 1897 and several others appeared later. As an example we will give Russell's paradox (1905), which can be presented using only the primary concepts of set theory and at the same time not violating the requirements of mathematical strictness. This is the paradox. Let us define M as the set of all those sets which do not contain themselves as an element. It would seem that this is an entirely proper definition because the formation of sets from sets is one of the bases of Cantor's theory. However, it leads to a contradiction. In order to make this

[4] Bourbaki, first essay, section "Set Theory."

clearer we shall use $P(x)$ to signify the property of set X of being an element of itself. In symbolic form this will be

$$P(x) \equiv x \epsilon x \qquad (1)$$

Then, according to the definition of set M, all its elements X have the property which is the opposite of $P(x)$:

$$x \epsilon M \equiv -P(x) \qquad (2)$$

Then we put the question: is set M itself an element, that is, is $P(M)$ true? If $P(M)$ is true, then $M \epsilon M$ according to definition (1). But in this case, substituting M for X in proposition (2) we receive $-P(M)$, for if M is included in set M, then according to the definition of the latter it should not have property P. On the other hand, if $P(M)$ is false, then $-P(M)$ occurs; then according to (2) M should be included in M, that is, $P(M)$ is true. Thus, $P(M)$ cannot be either true or false. From the point of view of formal logic we have proved two implications:

$$P(M) \supset -P(M)$$
$$-P(M) \supset P(M)$$

If the implication is expressed through negation and disjunction and we use the property of disjunction $A \lor A = A$, the first statement will become $-P(M)$ while the second will become $P(M)$. Therefore, a formal contradiction takes place and therefore anything you like may be deduced from set theory!

The paradoxes threatened set theory and the mathematical analysis based on it. Several philosophical-mathematical schools emerged which proposed different ways out of this blind alley. The most radical school was headed by Brouwer and came to be called *intuitionism;* this school demanded not only a complete rejection of Cantor's set theory, but also a radical revision of logic. Intuitionist mathematics proved quite complex and difficult to develop, and because it threw classical analysis onto the scrap heap most mathematicians found this position unacceptable. "No one can drive us from the heaven which Cantor created for us," Hilbert announced, and he found a solution which kept the basic content of set theory and at the same time eliminated the paradoxes and contradictions. With his followers Hilbert formulated the main channel along which the current of mathematical thought flowed.

Hilbert's solution corresponds entirely to the spirit of development of European mathematics. Whereas Cantor viewed his theory from a profoundly Platonist standpoint, as an investigation of the attributes of really existing and actually infinite sets, according to Hilbert the sets must be viewed as simply certain objects that satisfy axioms, while the axioms must be formulated so that definitions leading to paradoxes become impossible. The first system of set theory axioms which did not give rise to contradictions was proposed in 1908 by Zermelo and later modified. Other systems were also proposed, but the attitude toward set theory remained unchanged. In modern mathematics set theory plays the role of the frame, the skeleton which joins all its parts into a single whole but cannot be seen from the outside and does not come in direct contact with the external world. This situation can be truly understood and the formal and contentual aspects of mathematics combined only from the "linguistic" point of view regarding mathematics. This point of view, which we have followed persistently throughout this book, leads to the following conception. There are no actually infinite sets in reality or in our imagination. The only thing we can find in our imagination is the notion of potential infinity—that is, the possibility of repeating a certain act without limitation. Here we must agree fully with the intuitionist criticism of Cantor's set theory and give due credit to its insight and profundity. To use set theory in the way it is used by modern mathematics, however, it is not at all necessary to force one's imagination and try to picture actual infinity. The "sets" which are used in mathematics are simply symbols, linguistic objects used to construct models of reality. The postulated attributes of these objects correspond partially to intuitive concepts of aggregateness and potential infinity; therefore intuition helps to some extent in the development of set theory, but sometimes it also deceives. Each new mathematical (linguistic) object is defined as a "set" constructed in some particular way. This definition has no significance for relating the object to the external world, that is for interpreting it; it is needed only to coordinate it with the frame of mathematics, to mesh the internal wheels of mathematical models. So the language of set theory is in fact a metalanguage in relation to the language of contentual mathematics, and in this respect it is similar to the language of logic. If logic is the theory of proving mathematical statements, then set theory is the theory of constructing mathematical linguistic objects.

Precisely why did the intuitive concept of the set form the basis of mathematical construction? To define a newly introduced mathematical object means to point out its semantic ties with objects introduced before. With the exception of the trivial case where we are talking about redesignation, replacing a sign with a sign, there are always many such ties, and many objects introduced earlier can participate in them. And so, instead of saying that the new object is related in such-and-such ways to such-and-such old objects it is said that the new object is a set constructed of the old objects in such-and-such a manner. For example, a rational number is the result of dividing two natural numbers: the numerator by the denominator. The number $^5/_7$ is object X such that the value of the function "numerator" (X) is 5 and the value of the function "denominator" (X) is 7. In mathematics, however, the rational number is defined simply as a *pair* of natural numbers. In exactly the same way it would be necessary to speak only of the *realization* of a real number by different sequences of rational numbers, understanding this to mean a definite semantic relation between the new and old linguistic objects. Instead of this, it is said that the real number *is a set* of sequences of rational numbers. At the present time the terminology should be considered a vestige of Platonic views according to which what is important is not the linguistic objects but the elements of "ideal reality" concealed behind them, and therefore an object must be defined as a "real" set to acquire the right to exist. The idea of the set was promoted to "executive work" in mathematics as one of the aspects of the relation of name and meaning (specifically, that the meaning is usually a construction which includes *a number of* elements), and it is hardly necessary to prove that the relation of name and meaning always has been and always will be the basis of linguistic construction.

■ BOURBAKI'S TREATISE

AT THE CONCLUSION of this chapter we cannot help saying a few words about Bourbaki's multivolume treatise entitled *Éléments de mathematique*. Nicholas Bourbaki is a collective pseudonym used by a group of prominent mathematicians, primarily French, who joined together in the 1930s. *Éléments de mathematique* started publication in 1939.

Specialists from different fields of mathematics joined together in the Bourbaki group on the basis of a conception of mathematics as a formalized language. The goal of the treatise was to present all the most important achievements of mathematics from this point of view and to represent mathematics as one formalized language. And although Bourbaki's treatise has been criticized by some mathematicians for various reasons, it is unquestionably an important milestone in the development of mathematics along the path of self-awareness.

Bourbaki's conception was set forth in layman's terms in the article "The Architecture of Mathematics." At the start of the article the author asks: is mathematics turning into a tower of Babel, into an accumulation of isolated disciplines? Are we dealing with one mathematics or with several? The answer given to this question is as follows. Modern axiomatic mathematics is one formalized language that expresses abstract mathematical structures that are not distinct, independent objects but rather form a hierarchical system. By a "structure" Bourbaki means a certain number of relations among objects which possess definite properties. Leaving the objects completely undefined and formulating the properties of relations in the form of axioms and then extracting the consequences from them according to the rules of logical inference, we obtain an axiomatic theory of the given structure. Translated into our language, a structure is the semantic aspect of a mathematical model. Several types of fundamental generating structures may be identified. Among them are algebraic structures (which reflect the properties of the composition of objects), structures of order, and topological structures (properties related to the concepts of contiguity, limit, and continuity). In addition to the most general structure of the given type—that is, the structure with the smallest number of axioms—we find in each type of generating structure structures obtained by including additional axioms. Thus, group theory includes the theory of finite groups, the theory of abelian groups, and the theory of finite abelian groups. Combining generating structures produces complex structures such as, for example, topological algebra. In this way a hierarchy of structures emerges.

How is the axiomatic method employed in creative mathematics? This is where, Bourbaki writes, the axiomatic method is closest to the experimental method. Following Descartes, it "divides difficulties in order to resolve them better." In proofs of a complex theory it tries to break down the main groups of arguments involved and, taking them separately, deduce consequences from them (the dismemberment of models or structures, which we discussed above). Then, returning to the initial theory, it again combines the structures which have been identified beforehand and studies how they interact with one another. We conclude with this citation:

> From the axiomatic point of view, mathematics appears thus as a storehouse of abstract forms—the mathematical structures; and it so happens—without our knowing why—that certain aspects of empirical reality fit themselves into these forms, as if through a kind of preadaptation. Of course, it cannot be denied that most of these forms had originally a very definite intuitive content; but it is exactly by deliberately throwing out this content that it has been possible to give these forms all the power which they were capable of displaying and to prepare them for new interpretations and for the development of their full power.[5]

[5] Bourbaki, "The Architecture of Mathematics."

CHAPTER THIRTEEN
Science and Metascience

■ EXPERIMENTAL PHYSICS

WHEN THE FOUNDATIONS of the new mathematics were being constructed, at the turn of the seventeenth century, the basic principles of experimental physics were also developed. Galileo (1564–1642) played a leading role in this process. He not only made numerous discoveries and inventions which constituted an epoch in themselves, but also—in his books, letters, and conversations—taught his contemporaries a new method of acquiring knowledge. Galileo's influence on the minds of others was enormous. Francis Bacon (1566–1626) was also important in establishing experimental science. He gave a philosophical analysis of scientific knowledge and the inductive method.

Unlike the ancient Greeks, the European scientists were by no means contemptuous of empirical knowledge and practical activity. At the same time they were full masters of the theoretical heritage of the Greeks and had already begun making their own discoveries. This combination engendered the new method. "Those who have treated of the sciences," Bacon writes,

> have been either empirics or dogmatical. The former like ants only heap up and use their store, the latter like spiders spin out their own webs. The bee, a mean between both, extracts matter from the flowers of the garden and the field, but works and fashions it by its own efforts. The true labor of philosophy resembles hers, for it neither relies

entirely nor principally on the powers of the mind, nor yet lays up in the memory the matter afforded by the experiments of natural history and mechanics in its raw state, but changes and works it in the understanding. We have good reason, therefore, to derive hope from a closer and purer alliance of these faculties (the experimental and rational) than has yet been attempted.[1]

■ THE SCIENTIFIC METHOD

THE CONCEPT of the experiment assumes the existence of a theory. Without a theory there is no experiment; there is only observation. From the cybernetic (systems) point of view the experiment is a controlled observation; the controlling system is the scientific method, which relies on theory and dictates the organization of the experiment. Thus, the transition from simple observation to the experiment is a metasystem transition in the realm of experience and it is the first aspect of the emergence of the scientific method. Its second aspect is awareness of the scientific method as something standing above the theory—in other words, mastering the general principle of describing reality by means of formalized language, which we discussed in the previous chapter. As a whole, the emergence of the scientific method is one metasystem transition which creates a new level of control including control of observation (organization of the experiment) and control of language (development of theory). The new metasystem is what we mean by science in the modern sense of the word. Close direct and feedback ties are established between the experiment and the theory within this metasystem. Bacon describes them this way: "Our course and method . . . are such as not to deduce effects from effects, nor experiments from experiments (as the empirics do), but in our capacity of legitimate interpreters of nature, to deduce causes and axioms from effects and experiments."[2]

We can now give a final answer to the question: what happened in Europe in the early seventeenth century? A very major metasystem

[1] Francis Bacon, *Novum Organum*, Great Books of the Western World, Encyclopedia Britannica, 1955, Aphorism 95, p 126.
[2] Bacon, Aphorism 117, p 131.

transition took place, engulfing both linguistic and nonlinguistic activity. In the sphere of nonlinguistic activity it took shape as the experimental method. In the realm of linguistic activity it gave rise to the new mathematics, which has developed by metasystem transitions (the stairway effect) in the direction of ever-deeper self-awareness as a formalized language used to create models of reality. We described this process in the preceding chapter without going beyond mathematics. We can now complete this description by showing the system within which this process becomes possible. This system is science as a whole with the scientific method as its control device—that is, the aggregate of all human beings engaged in science who have mastered the scientific method together with all the objects used by them. When we were introducing the concept of the stairway effect in chapter 5 we pointed out that it takes place in the case where there is a metasystem Y which continues to be a metasystem in relation to systems of the series $X, X', X'', \ldots$, where each successive system is formed by a metasystem transition from the preceding one and, while remaining a metasystem, at the same time insures the possibility of metasystem transitions of smaller scale from X to X', from X'' to X'', and so on. Such a system Y possesses inner potential for development; we called it an *ultrametasystem*. In the development of physical production ultrametasystem Y is the aggregate of human beings who have the ability to convert means of labor into objects of labor. In the development of the exact sciences ultrametasystem Y is the aggregate of people who have mastered the scientific method—that is, who have the ability to create models of reality using formalized language.

We have seen that in Descartes the scientific method, taken in its linguistic aspect, served as a lever for the reform of mathematics. But Descartes did not just reform mathematics; while developing the same aspect of the same scientific method he created a set of theoretical models or hypotheses to explain physical, cosmic, and biological phenomena. If Galileo may be called the founder of experimental physics and Bacon its ideologist, then Descartes was both the founder and ideologist of theoretical physics. It is true that Descartes' models were purely mechanical (there could be no other models at that time) and imperfect, and most of them soon became obsolete. But those imperfections are not so important as the fact that Descartes established the principle of constructing theoretical models. In the nineteenth century, when the first knowledge of physics was accumulated and the mathematical apparatus was refined, this principle demonstrated its full utility.

It will not be possible here to give even a brief survey of the evolution of the ideas of physics and its achievements or the ideas and achievements of the other natural sciences. We shall dwell on two aspects of the scientific method which are universally important, namely the role of general principles in science and the criteria for selecting scientific theories, and then we shall consider certain consequences of the advances of modern physics in light of their great importance for the entire system of science and for our overall view of the world. At the conclusion of this chapter we shall discuss some prospects for the development of the scientific method.

■ THE ROLE OF GENERAL PRINCIPLES

BACON SET FORTH a program of gradual introduction of more and more general statements ("causes and axioms") beginning with unique empirical data. He called this process *induction* (that is to say, *introduction*) as distinguished from *deduction* of less general theoretical statements from more general principles. Bacon was a great opponent of general principles; he said that the mind does not need wings to raise it aloft, but lead to hold it on the ground. During the period of the "initial accumulation" of empirical facts and very simple empirical rules this conception still had some justification (it was also a counterbalance to Medieval Scholasticism), but it turned out later that the mind still needs wings more than lead. In any case, that is true in theoretical physics. To confirm this let us turn to Albert Einstein. In his article entitled "The Principles of Theoretical Physics," he writes:

> To apply his method the theoretician needs a foundation of certain general assumptions, so-called principles, from which he can deduce consequences. His activity thus breaks into two stages. In the first

place he must search for the principles, and in the second place he must develop the consequences which follow from these principles. School has given him good weapons to perform the second task. Therefore, if the first task has been accomplished for a certain area, that is to say a certain aggregate of interdependencies, the consequences will not be long in coming. The first task mentioned, establishing the principles which can serve as the basis for deduction, is categorically different. Here there is no method which can be taught and systematically applied to achieve the goal. What the investigator must do is more like finding in nature precisely formulated general principles which reflect definite general characteristics of the set of experimentally determined facts.[3]

In another article entitled "Physics and Reality," [4] Einstein speaks very categorically: "Physics is a devoloping logical system of thinking whose foundations cannot be obtained by extraction from past experience according to some inductive methods, but come only by free fantasy." The words about "free fantasy" do not mean, of course, that general principles do not depend on experience at all but rather that they are not determined uniquely by experience. The example Einstein often gave is that Newton's celestial mechanics and Einstein's general theory of relativity were constructed from the same facts of experience. But they began from completely different (in a certain sense even diametrically opposed) general principles, which is also seen in their different mathematical apparatuses.

As long as the edifice of theoretical physics had just a few "stories" and the consequences of general principles could be deduced easily and unambiguously, people were not aware that they had a certain freedom in establishing the principles. The distance between the trial and the error (or the success) in the trial and error method was so slight that they did not notice that they were using this method, but rather thought that they were deducing (although it was called inducing, not deducing) principles directly from experience. Einstein writes: "Newton, the creator of the first vast, productive

[3] See the collection A. Einstein, *Fizika i real'nost'* (Physics and Reality), Moscow: Nauka Publishing House, 1965. The quotations below are also taken from this Russian work. [Original article available in *Mein Weltbild*, Amsterdam, 1934—trans.]
[4] Original article in *Franklin Institute Journal*, vol 221, 1936, pp. 313–57—trans.

system of theoretical physics still thought that the basic concepts and principles of his theory followed from experience. Apparently this is how his statement, 'Hypotheses non fingo' (I do not compose hypotheses) must be understood.'' With time, however, theoretical physics changed into a multistory construction and the deduction of consequences from general principles became a complex and not always unambiguous business, for it often proved necessary in the process of deduction to make additional assumptions, most frequently "unprincipled" simplifications without which the reduction to numerical calculation would have been impossible. Then it became clear that between the general principles of the theory and the facts permitting direct testing in experience there is a profound difference; the former are free constructions of human reason, while the latter are the raw material reason receives from nature. True, we should not overestimate the profundity of this difference. If we abstract from human affairs and strivings it will appear that the difference between theories and facts disappears; both are certain reflections or models of the reality outside human beings. The difference lies in the level at which the models originate. The facts, if they are completely "de-ideologized," are determined by the effect of the external world on the human nervous system which we are compelled (for the present) to consider a system that does not permit alteration, and therefore we relate to facts as the primary reality. Theories are models embodied in linguistic objects. They are entirely in our power and thus we can throw out one theory and replace it with another just as easily as we replace an obsolete tool with a more highly refined one.

Growth in the abstractness (construct quality) of the general principles of physical theories and their remoteness from the immediate facts of experience leads to a situation in which it becomes increasingly more difficult using the trial and error method to find a trial which has a chance of success. Reason begins to experience an acute need for wings to soar with, as Einstein too is saying. On the other hand, the increase in the distance between general principles and verifiable consequences makes the general principles invulnerable to experience within certain limits, which was also frequently pointed out by the classics of modern physics. Upon finding a discrepancy between the consequences of a theory and the experiment, the investigator faces two alternatives: look for the causes of the discrepany in the general principles of the theory or look for them somewhere between the principles and the concrete consequences. In view of the great value of general principles and the significant expenditures required to revise the theory as a whole, the second path is always tried first. If the deduction of consequences from the general principles can be modified so that they agree with the experiment, and if this is done in a sufficiently elegant manner, everyone is appeased and the problem is considered solved. But sometimes the modification very clearly appears to be a patch, and sometimes patches are even placed on top of patches and the theory begins to tear open at the seams; nonetheless, its deductions are in agreement with the data of experience and continue to have their predictive force. Then these questions arise: what attitude should be taken toward the general principles of such a theory? Should we try to replace them with some other principles? What point in the "patchwork" process, how much "patching," justifies discarding the old theory?

■ CRITERIA FOR THE SELECTION OF THEORIES

FIRST OF ALL let us note that a clear awareness of scientific theories as linguistic models of reality substantially lessens the impact of the competition between scientific theories and the the naive point of view (related to Platonism) according to which the linguistic objects of a theory only express some certain reality, and therefore each theory is either "really" true if this reality actually exists or "really" false if this reality is fabricated. This point of view is engendered by transferring the status of the language of concrete facts to the language of concept-constructs. When we compare two competing statements such as "There is pure alcohol in this glass" and "There is pure water in this glass," we know that these statements permit an experimental check and that the one which is not confirmed loses all meaning as a model and all truth value. It is in fact false and only false. Things are entirely different with statements which express the general principles of scientific theories. Many verifiable consequences are deduced from them and if some of these prove false it is customary to say that the initial principles (or methods of deducing consequences) are not applicable to the given sphere of experience; it is usually possible to establish formal criteria of applicability. In a certain sense, therefore, general principles are "always true"; to be more precise, the concepts of truth and falsehood are not applicable to them, but the concept of their greater or lesser utility for describing real facts is applicable. Like the axioms of mathematics, the general principles of physics are abstract forms into which we attempt to squeeze natural phenomena. Competing principles stand out by how well they permit this to be done. But what does "well" mean?

If a theory is a model of reality, then obviously it is better if its sphere of application is broader and if it can make more predictions. Thus, the criterion of the generality and predictive power of a theory is the primary one for comparing theories. A second criterion is simplicity; because theories are models intended for use by people they are obviously better when they are simpler to use.

If scientific theories were viewed as something stable, not subject to elaboration and improvement, it would perhaps be difficult to suggest any other criteria. But the human race is continuously elaborating and improving its theories, which gives rise to one more criterion, the dynamic criterion, which is also the decisive one. In *The Philosophy of Science* this criterion was well stated by Phillip Frank:

> If we investigate which theories have actually been preferred because of their simplicity, we find that the decisive reason for acceptance has been neither economic nor esthetic, but rather what has often been called "dynamic." This means that the theory was preferred that proved to make science more "dynamic," i.e., more fit to expand into unknown territory. This can be made clear by using an example that we have invoked frequently in this book: the struggle between the Copernican and the Ptolemaic systems. In the period between Copernicus and Newton a great many reasons had been invoked on behalf of one or the other system. Eventually, however, Newton advanced his theory of motion, which accounted excellently for all motions of celestial bodies (e.g., comets), while Copernicus as well as Ptolemy had accounted for only the motions in our planetary system. Even in this re-

stricted domain, they neglected the "perturbations" that are due to the interactions between the planets. However, Newton's laws originated in generalizations of the Copernican theory, and we can hardly imagine how they could have been formulated if he had started with the Ptolemaic system. In this respect and in many others, the Copernican theory was the more "dynamic" one or, in other words, had the greater heuristic value. We can say that the Copernican theory was mathematically "simpler" and also more dynamic than the Ptolemaic theory.[5]

The esthetic criterion or the criterion of the beauty of a theory, which is mentioned by Frank, is difficult to defend as one independent of other criteria. But it becomes very important as an intuitive synthesis of all the above-mentioned criteria. To a scientist a theory seems beautiful if it is sufficiently general and simple and he feels that it will prove to be dynamic. Of course, he may be wrong in this too.

■ THE PHYSICS OF THE MICROWORLD

IN BOTH PHYSICS and pure mathematics, as the abstractness of the theories increased the understanding of their linguistic nature became solidly rooted. The decisive impetus was given to this process in the early twentieth century when physics entered the world of atoms and elementary particles, and quantum mechanics and the theory of relativity were created. Quantum mechanics played a particularly large part. This theory cannot be understood at all unless one constantly recalls that it is just a linguistic model of the microworld, not a representation of how it would "really" look if it were possible to see it through a microscope with monstrous powers of magnification; there is no such representation nor can there be one. Therefore the notion of the theory as a linguistic model of reality became a constituent part of modern physics, essential for successful work by physicists. Consequently their attitude toward the nature of their work also began to change. Formerly the theoretical physicist felt himself to be the discoverer of something which existed before him and was independent

[5] Phillip Frank, *The Philosophy of Science* (Englewood Cliffs, New Jersey: Prentice-Hall, 1957).

of him, like a navigator discovering new lands; now he feels he is more a creator of something new, like a master artisan who creates new buildings, machines, and tools and has complete mastery of his own tools. This change has even appeared in our way of talking. Traditionally, Newton is said to have "discovered" [otkryl] infinitesimal calculus and celestial mechanics; when we speak of a scientist today we say that he has "created" [sozdal], "proposed" [predlozhil], or "worked out" [razrabotal] a new theory. The expression "discovered" sounds archaic. Of course, this in no way diminishes the merits of the theoreticians, for creation is as honorable and inspiring an occupation as discovery.

But why did quantum mechanics require awareness of the "linguistic quality" of theories?

According to the initial atomistic conception, atoms were simply very small particles of matter, small corpuscles which had, in particular, a definite color and shape which determined the color and physical properties of larger accumulations of atoms. The atomic physics of the early twenth century transferred the concept of indivisibility from the atom to elementary particles—the electron, the proton, and soon after the neutron. The word "atom" began to mean a construction consisting of an atomic nucleus (according to the initial hypothesis it had been an accumulation of protons and electrons) around which electrons revolved like planets around the sun. This representation of the structure of matter was considered hypothetical but extremely plausible. The hypothetical quality was understood in the sense discussed above: the planetary model of the atom must be either true or false. If it is true (and there was virtually no doubt of this) then the electrons "really" are small particles of matter which describe certain trajectories around a nucleus. Of course, in comparison with the atoms of the ancients, the elementary particles were already beginning to lose some properties which would seem to be absolutely essential for particles of matter. It became clear that the concept of color had absolutely no application to electrons and protons. It was not that we did not know what color they were; the question was simply meaningless, for color is the result of interaction with light by at least the whole atom, and more precisely by an accumulation of many atoms. Doubts also arose regarding the concepts of the shape and dimensions of electrons. But the most sacred element of the representation of the material particle, that the particle has a definite position in space at each moment, remained undoubted and taken for granted.

■ THE UNCERTAINTY RELATION

QUANTUM MECHANICS destroyed this notion, through the force of new experimental data. It turned out that under certain conditions elementary particles behave like waves, not particles; in this case they are not "blurred" over a large area of space, but keep their small dimensions and discreteness. The only thing that is blurred is the probability of finding them at a particular point in space.

As an illustration of this let us consider figure 13.1. The figure shows an electron gun which sends electrons at a certain velocity toward a diaphragm behind which stands a screen. The diaphragm is made of a material which is impervious to electrons, but it has two holes through which electrons pass to strike the screen. The screen is coated with a substance that fluoresces when acted upon by electrons, so that there is a flash at the place struck by an electron. The stream of electrons from the gun is sufficiently infrequent so that each electron passes through the diaphragm and is recorded on the screen independently of others. The distance between the holes in the diaphragm is many times greater than the dimensions of the electrons

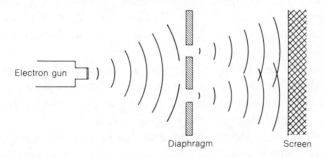

Figure 13.1. Diffraction of electrons.

(according to any estimate of their size) but comparable with the quantity h/p where h is the Planck constant and p is the momentum of the electron—i.e., the product of its velocity and mass.

These are the conditions of the experiment. The result of the experiment is a distribution of flashes on the screen. The first conclusion from analyzing the results of the experiment is the following: electrons strike different points of the screen and it is impossible to predict which point each electron will strike. The only thing that can be predicted is the probability that a particular electron will strike a particular point—that is, the average density of flashes after a very large number of electrons have struck the screen. But this is just half the trouble. One can imagine that different electrons pass through different parts of the hole in the diaphragm, experience effects of differing force from the edges of the holes, and therefore are deflected differently. The real troubles arise when we begin to investigate the average density of flashes on the screen and compare it with the results which are obtained when we close one of the holes in the diaphragm. If an electron is a small particle of matter, then when it reaches the region of the diaphragm it is either absorbed or passes through one of the holes. Because the holes in the diaphragm are set symmetrically relative to the electron gun, on the average half of the electrons pass through each hole. This means that if we close one hole and pass 1 million electrons through the diaphragm then close the second hole and open the first and pass 1 million more electrons through, we should receive the same average density of flashes as if we were to pass 2 million electrons through the diaphragm with two holes open. But it turns out that this is not the case! With two holes open the distribution is different; it contains maximums and minimums as is the case in diffraction of waves.

The average denisty of flashes can be calculated by means of quantum mechanics, relating the electrons to the so-called *wave function,* which is a certain imaginary field whose intensity is proportional to the probability of the observed events.

It would take too much space to describe all the attempts, none successful, which have been made to correlate the representation of the electron as a "conventional" particle (such particles have come to be called *classical*, as opposed to *quantum* particles) with the experimental data on electron behavior. There is a vast literature, both specialized and popular, devoted to this question. The following two things have become clear. In the first place, if we simultaneously measure the coordinate of a quantum particle (any such particle, not necessarily an electron) on a certain axis X and the momentum in this direction p, the errors of measurement, which we designate Δ_x and Δ_p respectively, comply with Heisenberg's uncertainty relation:

$$\Delta_x \Delta_p \geqslant h$$

No clever tricks can get around this relation. When we try to measure coordinate X more exactly the spread of magnitudes of momentum p is larger, and vice versa. The uncertainty relation is a universally true law of nature, but because the Planck constant h is very small, the relation plays no part in measurements of bodies of macroscopic size.

In the second place, the notion that quantum particles *really* move along certain completely definite trajectories—which is to say at each moment they *really* have a completely definite coordinate and velocity (and therefore also momentum) which we are simply unable to measure exactly—runs up against insurmountable logical difficulties. On the other hand, the refusal on principle to ascribe a real trajectory to the quantum particle and adoption of the tenet that the most complete description of the state of a particle is an indication of its wave function yields a logically flawless, mathematically simple and elegant theory which fits brilliantly with experimental facts; specifically, the uncertainty relation follows from it immediately. This is the theory of quantum mechanics. The work of Niels Bohr (1885–1962) the greatest scientist-philosopher of our time, played the major part in clarifying the physical and logical foundations of quantum mechanics and interpreting it philosophically.

■ GRAPHIC AND SYMBOLIC MODELS

SO AN ELECTRON does not have a trajectory. The most that can be said of an electron is an indication of its wave function whose square will give us the probability of finding the electron in the proximity of a particular point in space. But at the same time we say that the electron is a material particle of definite (and very small) dimensions. Combining these two representations, as was demanded by observed facts, proved a very difficult matter and even today there are still people who reject the standard interpretation of quantum mechanics (which has been adopted by a large majority of physicists following the Bohr school) and want to give the quantum particles back their trajectories no matter what. Where does such persistence come from? After all, the expropriation of color from the electrons was completely painless and, from a logical point of view, recognizing that the concept of trajectory cannot apply to the electron is no different in principle from recognizing that the concept of color does not apply. The difference here is that when we reject the concept of color we are being a little bit hypocritical. We say that the electron has no color, but we ourselves picture it as a little greyish (or shiny, it is a matter of taste) sphere. We substitute an *arbitrary* color for the *absence* of color and this does not hinder us at all in using our model. But this trick does not work in relation to position in space. The notion of an electron which is located somewhere at every moment hinders understanding of quantum mechanics and comes into contradiction with experimental data. Here we are forced to reject completely the graphic-geometric representation of particle movement. And this is what causes the painful reaction. We are so accustomed to associating the space-time picture with true reality, with what exists objectively and independently of us, that it is very difficult for us to believe in an objective reality which does not fit within this conception. And we ask ourselves again and again: after all, if the electron is not ''blurred'' in space, then it must *really* be somewhere, mustn't it?

It requires real mental effort to recognize and feel the meaninglessness of this question. First we must be aware that all our knowledge and theories are secondary models of reality, that is, models of the primary models which are the data of sensory experience. These data bear the ineradicable imprint of the organization of our nervous system and because space-time concepts are set in the very lowest levels of the nervous system, none of our perceptions and representations, none of the products of our imagination, can go outside the framework of space-time pictures. But this framework can still be broadened to some extent. This must be done, however, not by an illusory movement "downward," toward objective reality "as it is, independent of our sense organs," but rather by a movement "upward," that is, by constructing secondary symbolic models of reality. Needless to say, the symbols of the theory preserve their continuous space-time existence just as the primary data of experience do. But in the relations between the one and the other, which is to say in the semantics of the theory, we can allow ourselves significant freedom if we are guided by the logic of new experimental facts, and not by our customary space-time intuition. And we can construct a sign system whose functioning is in no way related to graphic representations but is entirely appropriate to the condition of adequately describing reality. Quantum mechanics is such a system. In this system the quantum particle is neither a little greyish sphere nor a shiny one, and it is not a geometric point; it is a certain concept, a functional node of the system which, together with the other nodes, ensures description and anticipation of the real facts of experience: flashes on the screen, instrument readings, and the like.

Let us return to the question of how the electron "really" moves. We have seen that, owing to the uncertainty relation, the experiment cannot in principle give an answer to this question. This question is therefore meaningless as an "external part" of the physical model of reality. All that we can do is to ascribe a purely theoretical meaning to it. But then it loses its direct linkage with observed phenomena and the expression "really" becomes pure deception! When we go outside the sphere of perception and declare that such-and-such "really" takes place we are always moving upward, not downward; we are constructing a pyramid of linguistic objects and it is only because of the optical illusion that it seems to us we are going deeper into the realm which lies beneath sensory experience. To put it metaphorically, the plane that separates sensory experience from reality is absolutely impervious; and when we attempt to discern what is going on beneath it we see only the upside-down reflection of the pyramid of theories. This does not mean that true reality is unknowable and our theories are not correct models of it; one must remember, however, that all these models lie on this side of sensory experience and it is meaningless to correlate distinct elements of theories with the illusory "realities" on the other side, as was done by Plato for example. The representation of the electron as a little sphere moving along a trajectory is just as much a construction as is the interlinking of the symbols of quantum theory. It differs only in that it includes a space-time picture to which, following convention, we ascribe illusory reality by using the expression "really," which is meaningless in this case.

The transition to conscious construction of symbolic models of reality that do not rely on any graphic representations of physical objects is the great philosophical achievement of quantum mechanics. In fact physics has been a symbolic model since Newton's time and it owes its successes (numerical calculations) to precisely this symbolic nature; but graphic representations were present as an essential element. Now they are not essential and this has broadened the class of possible models. Those who want to bring back the graphic quality no matter what, although they see that the theory works better without it, are in fact asking that the class of models be narrowed. They will hardly be successful. They are like the odd fellow who hitched his horse to a steam locomotive for, although he could see that the train moved without a horse, it was beyond his powers to recognize such a situation as normal. Symbolic models are a steam engine which has no need to be harnessed to the horse of graphic representations for each and every concept.

■ THE COLLAPSE OF DETERMINISM

THE SECOND IMPORTANT result of quantum mechanics, the collapse of determinism, was significant in general philosophy. Determinism is a philosophical concept. It is the name used for the view which holds that all events occurring in the world have definite causes and necessarily occur; that is, they cannot not occur. Attempts to make this definition more precise reveal the logical defects in it which hinder precise formulation of this viewpoint as a scientific proposition without introducing any additional representations about objective reality. In fact, what does "events have causes" mean? Can it really be possible to indicate some finite number of "causes" of a given event and say that there are no others? And what does it mean that the event "cannot not occur?" If this means only that it has occurred then the statement becomes a tautology.

Philosophical determinism can, however, obtain a more precise interpretation within the framework of a scientific theory which claims to be a universal description of reality. It actually did receive such an interpretation within the framework of *mechanism* (mechanical philosophy), the philosophical-scientific conception which emerged on the basis of the advances of classical mechanics in application to the motions of the celestial bodies. According to the mechanistic conception the world is three-dimensional Euclidean space filled with a multitude of elementary particles which move along certain trajectories. Forces operate among the particles depending on their arrangement relative to one another and the movement of particles follows the laws of Newton's mechanics. With this representation of the world, its exact state (that is, the coordinates and velocities of all particles) at a certain fixed moment in time uniquely determines the exact state of the world at any other moment. The famous French mathematician and astronomer P. Laplace (1749–1827) expressed this proposition in the following words:

> Given for one instance an intelligence which could comprehend all the forces by which nature is animated and the respective situation of the beings who compose it—an intelligence sufficiently vast to submit these data to analysis—it would embrace in the same formula the movements of the greatest bodies of the universe and those of the lightest atom; for it, nothing would be uncertain and the future, as the past, would be present to its eyes.[6]

This conception became called *Laplacian determinism*. It is a proper and inevitable consequence of the mechanistic conception of the world. It is true that Laplace's formulation requires a certain refinement from a modern point of view because we cannot recognize

[6] Laplace, P., *Opyt filosofii terorii veroyatnostei*, Moscow, 1908, p 9. [Original *Essai philosphique des probabilités*, 1814. English translation, *A Philosophical Essay on Probabilities*, New York: Dover Publications, 1951—trans.]

as proper the concepts of an all-knowing reason or absolute precision of measurement. But it can be modernized easily, almost without changing its meaning. We say that if the coordinates and velocities of all particles in a sufficiently large volume of space are known with adequate precision then it is possible to calculate the behavior of any system in any given time interval with any given precision. The conclusion that all future states of the universe are predetermined can be drawn from this formulation just as from Laplace's initial formulation. By unrestrictedly increasing the precision and scope of measurements we unrestrictedly extend prediction periods. Because there are no restrictions in principle on the precision and range of measurements (that is, restrictions which follow not from the limitations of human capabilities but from the nature of the objects of measurement) we can picture the extreme case and say that really the entire future of the world is already absolutely and uniquely determined today. In this case the expression "really" acquires a perfectly clear meaning; our intuition easily recognizes that this "really" is proper and we object to its discrediting.

Thus, the mechanistic conception of the world leads to the notion of the complete determinism of phenomena. But this contradicts our own subjective feeling of free choice. There are two ways out of this: to recognize the feeling of freedom of choice as "illusory" or to recognize the mechanistic conception as unsuitable as a universal picture of the world. It is already difficult today to say how thinking people of the "pre-quantum" age were divided between these two points of view. If we approach the question from a modern standpoint, even knowing nothing of quantum mechanics, we must firmly adhere to the second point of view. We now understand that the mechanistic conception, like any other conception, is only a secondary model of the world in relation to the primary data of experience; therefore the immediate data of experience always have priority over any theory. The feeling of freedom of choice is a primary fact of experience just like other primary facts of spiritual and sensory experience. A theory cannot refute this fact; it can only correlate new facts with it, a procedure which, where certain conditions are met, we call *explanation* of the fact. To declare freedom of choice "illusory" is just as meaningless as telling a person with a toothache that his feeling is "illusory." The tooth may be entirely healthy and the feeling of pain may be a result of stimulation of a certain segment of the brain, but this does not make it "illusory."

Quantum mechanics destroyed determinism. Above all the representation of elementary particles as little corpuscles moving along definite trajectories proved false, and as a consequence the entire mechanistic picture of the world—which was so understandable, customary, and seemingly absolutely beyond doubt—also collapsed. Twentieth-century physicists can no longer tell people what the world in which they live is *really* like, as nineteenth-century physicists could. But determinism collapsed not only as a part of the mechanistic conception, but also as a part of any picture of the world. In principle one could conceive of a complete description (picture) of the world that would include only really observed phenomena but would give unambiguous predictions of all phenomena that will ever be observed. We now know that this is impossible. We know situations exist in which it is impossible in principle to predict which of the sets of conceivable phenomena will actually occur. Moreover, according to quantum mechanics these situations are not the exception; they are the general rule. Strictly determined outcomes are the exception to the rule. The quantum mechanics description of reality is a fundamentally probabilistic description and includes unequivocal predictions only as the extreme case.

As an example let us again consider the experiment with electron diffraction depicted in figure 13.1. The conditions of the experiment are completely determined when all geometric parameters of the device and the initial momentum of the electrons released by the gun are given. All the electrons propelled from the gun and striking the screen are operating under the same conditions and are described by the same wave function. However, they are absorbed (produce flashes) at different points of the screen, and it is impossible to predict beforehand at what point an electron will produce a flash. It is even impossible to predict whether the electron will be deflected upward or downward in our picture; all that can be done is to indicate the probability of striking different segments of the screen.

It is permissible, however, to ask the following question: why are we confident that if quantum mechanics cannot predict the point which an electron will strike no other future theory will be able to do this?

We shall give two answers to this question. The first answer can be called formal. Quantum mechanics is based on the principle that description by means of the wave function is a maximally complete description of the state of the quantum particle. This principle, in the form of the uncertainty relation that follows from it, has been confirmed by an enormous number of experiments whose interpretation contains nothing but concepts of the lowest level, directly linked to observed quantities. The conclusions of quantum mechanics, including the more complex mathematical calculations, have been confirmed by an even larger number of experiments. And there are absolutely no signs that we should doubt this principle. But this is equivalent to the impossibility of predicting the exact outcome of an experiment. For example, to indicate what point on the screen an electron will strike one must have more knowledge about it than the wave function provides.

The second answer requires an understanding of why we are so disinclined to agree that it is impossible to predict the point the electron will strike. Centuries of development in physics have accustomed people to the thought that the movement of inanimate bodies is controlled exclusively by causes external to them and that these causes can always be discovered by sufficiently precise investigation. This statement was completely justified as long as it was considered possible to watch a system without affecting it, which held true for experiments with macroscopic bodies. Imagine that figure 13.1 shows the distribution of cannonballs instead of electrons, and that we are studying their movement. We see that in one case the ball is deflected upward while in another it goes downward; we do not want to believe that this happens by itself, but are convinced that the difference in the behavior of the cannonballs can be explained by some real cause. We photograph the flight of the ball, do some other things, and finally find phenomena A_1 and A_2, which are linked to the flight of the cannonball in such a way that where A_1 is present the ball is deflected upward and where A_2 is present it goes downward. We therefore say that A_1 is the cause of deflection upward while A_2 is the cause of deflection downward. Possibly our experimental area will prove inadequate or we shall simply get tired of investigating and not find the sought-for cause. We shall still remain convinced that a cause really exists, and that if we had looked harder we would have found phenomena A_1 and A_2.

In the experiment with electrons, once again we see that the electron is deflected upward in some cases and downward in others and in the search for the cause we try to follow its movement, to peek behind it. But it turns out here that we cannot peek behind the electron without having a most catastrophic effect on its destiny. A stream of light must be directed at the electron if we are to "see" it. But the light interacts with the substance in portions, quanta, which obey the same uncertainty relation as do electrons and other particles. Therefore it is not possible to go beyond the uncertainty relation by means of light or by any other investigative means. In attempting to determine the coordinate of the electron more precisely by means of photons we either transfer such a large and indeterminate momentum to it that it spoils the entire experiment or we measure the coordinate so crudely that we do not find out anything new about it. Thus, phenomena A_1 and A_2 (the causes according to which the electron is deflected upward in some cases and downward in others) do not exist in reality. And the statement that there "really" is some cause loses any scientific meaning.

Thus, there are phenomena that have no causes, or more precisely, there are series of possibilities from which one is realized without any cause. This does not mean that the principle of causality should be entirely discarded; in the same experiment, by turning off the electron gun we cause the flashes on the screen to completely disappear, and turning off the gun does cause this. But this does mean that the principle must be narrowed considerably in comparison with the way it was understood in classical mechanics and the way it is still understood in the ordinary consciousness. Some phenomena have no causes; they must be accepted simply as something given. That is the kind of world we live in.

The second answer to the question about the reasons for our confidence that unpredictable phenomena exist is that the uncertainty relation assists us in clarifying not only a mass of new facts but also the nature of the break regarding causality and predictability that occurs when we enter the microworld. We see that belief in absolute causality originated from an unstated assumption that there are infinitely subtle means of watching and investigating, of "peeking" behind the object. But when they came to elementary particles physicists found that there is a minimum quantum of action measurable by the Planck constant h, and this creates a vicious circle in attempts to make the description of one particle by means of another detailed beyond measure. So absolute causality collapsed, and with it went determinism. From a general philosophical point of view it is entirely natural that if matter is not infinitely divisible then description cannot be infinitely detailed so that the collapse of determinism is more natural than its survival would have been.

■ "CRAZY" THEORIES AND METASCIENCE [7]

THE ABOVEMENTIONED SUCCESSES of quantum mechanics refer primarily to the description of nonrelativistic particles—that is, particles moving at velocities much slower than the velocity of light, so that effects related to relativity theory (relativistic effects) can be neglected. We had nonrelativistic quantum mechanics in mind when we spoke of its completeness and logical harmony. Nonrelativistic quantum mechanics is adequate to describe phenomena at the atomic level, but the physics of elementary high-energy particles demands the creation of a theory combining the ideas of quantum mechanics with the theory of relativity. Only partial successes have been achieved thus far on this path; no single, consistent theory of elementary particles which explains the enormous material accumulated by experimenters exists. Attempts to construct a new theory by superficial modifications of the old theory do not yield significant results. Creation of a satisfactory theory of elementary particles runs up

[7] This section is written on the motifs of the author's article published under the same title in the journal *Voprosy filosofii* (Questions of Philosophy), No 5, 1968.

against the uniqueness of this realm of phenomena, phenomena which seem to take place in a completely different world and demand for their explanation completely unconventional concepts which differ fundamentally from our customary scheme of concepts.

In the late 1950s Heisenberg proposed a new theory of elementary particles. Upon becoming familiar with it Bohr said that it could hardly prove true because it was "not crazy enough." The theory was not in fact recognized, but Bohr's pointed remark became known to all physicists and even entered popular writing. The word "crazy" [Russian *sumasshedshaya*, literally "gone out of the mind"] was naturally associated with the epithet "strange," which was applied to the world of elementary particles. But does "crazy" mean just "strange," "unusual"? Probably if Bohr had said "not unusual enough," it would not have become an aphorism. The word "crazy" has a connotation of "unreasoned," "coming from an unknown place," and brilliantly characterizes the current situation of the theory of elementary particles, in which everyone recognizes that the theory must be fundamentally revised, but no one knows how to do it.

The question arises: does the "strangeness" of the world of elementary particles—the fact that our intuition, developed in the macroworld, does not apply to it—doom us to wander eternally in the darkness?

Let us look into the nature of the difficulties which have arisen. The principle of creating formalized linguistic models of reality did not suffer in the transition to study of the microworld. But if the wheels of these models, the physical concepts, came basically from our everyday macroscopic experience and were only refined by formalization, then for the new, "strange" world we need new, "strange" concepts. But we have nowhere to take them from; they will have to be constructed and also combined properly into a whole scheme. In the first stage of study of the microworld the wave function of nonrelativistic quantum mechanics was constructed quite easily by relying on the already existing mathematical apparatus used to describe macroscopic phenomena (the mechanics of the material point, the mechanics of continuous media, and matrix theory). Physicists were simply lucky. They found prototypes of what they needed in two (completely different) concepts of macroscopic physics and they used them to make a "centaur," the quantum concept of the wave-particle. But we cannot count on luck all the time. The more deeply we go into the microworld the greater are the differences between the wanted concept-constructs and the ordinary concepts of our macroscopic experience; it thus becomes less and less probable that we shall be able to improvise them, without any tools, without any theory. Therefore we must subject the very task of constructing scientific concepts and theories to scientific analysis, that is, we must make the next metasystem transition. In order to construct a definite physical theory in a qualified manner we need a general theory of the construction of physical theories (a metatheory) in the light of which the way to solve our specific problem will become clear.

The metaphor of the graphic models of the old physics as a horse and the abstract symbolic models as a steam engine can be elaborated as follows. Horses were put at our disposal by nature. They grow and reproduce by themselves and it is not necessary to know their internal organization to make use of them. But we ourselves must build the steam engine. To do this we must understand the principles of its organization and the physical laws on which they are based and furthermore we must have certain tools for the work. In attempting to construct a theory of the "strange" world without a metatheory of physical theories we are like a person who has decided to build a steam engine with his bare hands or to build an airplane without having any idea of the laws of aerodynamics.

And so the time has come for the next metasystem transition. Physics needs . . . I want to say "metaphysics," but, fortunately for our terminology, the metatheory we need is a metatheory in relation to any natural science theory which has a high degree of formalization and therefore it is more correct to call it a *metascience*. This term has the shortcoming of creating the impression that a metascience is something fundamentally outside of science whereas in fact the new level of the hierarchy created by this metasystem transition must, of course, be included in the general body of science, thereby broadening it. The situation here is similar to the situation with the term metamathematics; after all, metamathematics is also a part of mathematics. Inasmuch as the term "metamathematics" was acceptable nonetheless, the term "metascience" may also be considered acceptable. But because a very important part of metascientific investigation is investigation of the concepts of a theory, the term *conceptology* may also be suggested.

The basic task of metascience can be formulated as follows. A certain aggregate of facts or a certain generator of facts is given. How can one construct a theory that describes these facts effectively and makes correct predictions?

If we want metascience to go beyond general statements it must be constructed as a full-fledged mathematical theory and its object, the natural science theory, must be presented in a formalized (albeit simplified; such is the price of formalization) manner, subject to mathematics. Represented in this form the scientific theory is a formalized linguistic model whose mechanism is the hierarchical system of concepts, a point of view we have carried through the entire book. From it, the creation of a mathematical metascience is the next natural metasystem transition, and when we make this transition we make our objects of study formalized languages as a whole—not just their syntax but also, and primarily, their semantics, their application to description of reality. The entire course of development of physico-mathematical science leads us to this step.

But in our reasoning thus far we have been basing ourselves on the needs of physics. How do things stand from the point of view of pure mathematics?

Whereas theoretical physicists know what they need but can do little, "pure" mathematicians might rather be reproached for doing a great deal but not knowing what they need. There is no question that many pure mathematical works are needed to give cohesion and harmony to the entire edifice of mathematics, and it would be silly to demand immediate "practical" application from every work. All the same, mathematics is created to learn about reality, not for esthetic or sporting purposes like chess, and even the highest stages of mathematics are in the last analysis needed only to the extent that they promote achievement of this goal.

Apparently, upward growth of the edifice of mathematics is always necessary and unquestionably valuable. But mathematics is also growing in breadth and it is becoming increasingly difficult to determine what is needed and what is not and, if it is needed, to what extent. Mathematical technique has now developed to the point where the construction of a few new mathematical objects within the framework of the axiomatic method and investigation of their characteristics has become almost as common, although not always as easy, a matter as computations with fractions were for the Ancient Egyptian scribes. But who knows whether these objects will prove necessary? The need is emerging for a theory of the application of mathematics, and this is actually a metascience. Therefore, the development of metascience is a guiding and organizing task in relation to the more concrete problems of mathematics.

The creation of an effective metascience is still far distant. It is difficult today to even picture its general outlines. Much more preparatory work must be done to clarify them. Physicists must master "Bourbakism" and develop a "feel" for the play of mathematical structures, which leads to the emergence of rich axiomatic theories suitable for detailed description of reality. Together with mathematicians they must learn to break symbolic models down into their individual elements of construction in order to compose the necessary blocks from them. And of course, there must be development of the technique of making formal computations with arbitrary symbolic expressions (and not just numbers) using computers. Just as the transition from arithmetic to algebra takes place only after complete assimilation of the technique of arithmetic computations, so also the transition to the theory of creating arbitrary sign systems demands highly sophisticated techniques for operations on symbolic expressions and a practical answer to the problem of carrying out cumbersome formal computations. Whether the new method will contribute to a resolution of the specific difficulties that now face the theory of elementary particles or whether they will be resolved earlier by "old-time" manual methods we do not know; and in the end it is not important because new difficulties will undoubtedly arise. One way or another, the creation of a metascience is on the agenda. Sooner or later it will be solved, and then people will receive a new weapon for conquering the strangest and most fantastic worlds.

CHAPTER FOURTEEN
The Phenomenon of Science

■ THE HIGHEST LEVEL OF THE HIERARCHY

THE UNIVERSE IS EVOLVING. The organization of matter is constantly growing more complex. This growing complexity occurs through metasystem transitions from which new levels of organization emerge which are levels of the control hierarchy. The inorganic world, plants, animals, the human being—such has been the course of evolution on our planet, and as far as we know this is the greatest advance which has been made in the part of space that surrounds us. It also seems highly probable that the human being is the crown of evolution of the entire cosmos. In any case, we do not have any direct indications or even the slightest hints of the existence of a higher level of organization. Therefore all we can do is consider ourselves the highest.

The appearance of the human being marks the beginning of the Age of Intellect, when the leading force of development becomes conscious human creativity and the highest level of organization is the culture of human society. In its development culture generates the next level of the hierarchy within itself. This is critical thinking which, in its turn, gives rise to modern science, constructing models of reality using sign systems. These are new models; they did not and could not exist in the minds of individual human beings outside of civilization and culture, and they enlarge human power over nature colossally. They make up the continuously improving and developing super-brain of the super-being which is humanity as a whole. Thus, science is the highest level of the hierarchy in the organization of cosmic matter. It is the highest growth point of a growing tree, the leading shoot in the evolution of the universe. This is the significance of the cosmic phenomenon of science as a part of the phenomenon of man.

■ SCIENCE AND PRODUCTION

JUST AS IN THE EVOLUTION of animals there was a stage when the central nervous system formed and as a result profound changes occurred in the structure, behavior, and external appearance of the organism, an age of swift and profound changes under the direct influence of science has now arrived in the development of society. At the beginning of the first industrial revolution science played a relatively small part, but then came discoveries in physics and chemistry which led to revolutionary changes in technology and the conditions of societal life. In the 1950s the second industrial revolution began, indebted entirely to scientific advances. It is still picking up speed today and even its very immediate repercussions are difficult to anticipate.

It is now widely recognized that science has become a direct productive force. On the other hand, it cannot develop without the development of industrial production, and that is becoming increasingly expensive. Modern production requires not only that ready formulas from science be used but also that scientific research and the scientific approach be introduced in all elements of production. More and more it comes to resemble science. On the other hand, science, attracting a significant part of the human and physical resources of society and becoming a regulated, mass occupation, is acquiring the characteristics of production. Science and production are growing together into a single hierarchical system. The uppermost growth point sends out leaves which grow rapidly at first but then stop and become standard, stable forms of interaction with physical reality: electrical motors, airplanes, machines to produce synthetic fabrics, and genetic methods of selection. But the growth point rises higher and higher and generates more and more new leaves.

■ THE GROWTH OF SCIENCE

SCIENCE IS GROWING. It grows exponentially, which is to say that its quantitative characteristics increase so many times each so many years. The total number of articles in scientific journals throughout the world doubles every 12 to 15 years.[1]

The number of workers in science doubles every 15 years in Western Europe, every 10 years in the United States, and every 7 years in the USSR. With such a furious growth rate the contemporary generation of scientists consitutes 90 percent of all the scientists who have ever lived on Earth.

Along with science other quantitative characteristics of the human race are growing exponentially: the total number of people and the total volume of production of material goods. But science significantly surpasses them in growth rate. The growth rates of population, production, and science are roughly in the ratio 1:2:4. This is a healthy ratio which reflects that evolution of an organism where the mass of muscles is growing more rapidly than the total mass of the body but the mass of the brain is growing more rapidly than the mass of the muscles. Unfortunately, the territorial distribution of growth is poor. High population growth falls primarily in countries with low production growth and virtually no contribution to world science. We hope, however, that humanity will be able to handle these growing pains. There can hardly be any doubt that growing pains is all they are. After all, the rapid population growth in the underdeveloped countries is due to the high level of world science (medical service, social changes). Already today the human race represents a highly integrated system and its overall takeoff, which is conveyed by the ratio 1:2:4, is the result of the development of science, a very recent phenomenon. If we extrapolate the present rate of population growth

[1] The figures are taken from G. N. Dobrov's book *Nauka o nauke* (The Science of Science), Kiev, 1966.

(on the order of two percent a year) into the past, it appears that there would have been just two people living on Earth a mere thousand years ago!

The proportion of people employed directly in the sphere of science is still small, even in the highly developed countries. It ranges from 0.5 to one percent. The figure is now growing rapidly, but it is obvious that sooner or later its growth will slow down; it will reach a constant level which is difficult to predict today. As far as can be judged by the literature, it is considered improbable that this level will exceed 25 percent. After all, by weight the human brain is also a small part of the entire body.

The absolute number of people engaged in scientific work will nonetheless grow steadily, and together with it the quantity of information produced by them will also grow steadily. This quantity is already enormous today. The first scientific periodicals began to come out in the second half of the seventeenth century. By the start of the 1960s the total number of periodicals was about 50,000 (see figure 14.1), 30,000 of which were still being published in 1966. A total of 6 million articles had been published in them, and this figure was increasing by 500,000 a year.[2] The total number of patents and author's certificates recorded was more than 13 million. This stream of information, which must be used, gives rise to serious difficulties. For a long time scientific work has demanded an extreme degree of specialization, but recently it has become increasingly common for scientists to be unable to follow all the new work even in their own narrow areas. They face a dilemma: either read articles or work. Moreover, as a result of technical difficulties in disseminating and processing enormous amounts of information (we might also mention the imperfections in the information system in science and technology) substantial effort must often be expended to find the necessary information, and this effort is not always successful. As a result a great deal of work is duplicated or not properly done. According to estimates by American scientists, between 10 and 20 percent of sci-

[2] The figures are taken from D. Price's "Little Science, Big Science," in the collection of articles *Nauka o nauke* (The Science of Science), Moscow, Progress Publishing House, 1966; original: Columbia University Press, 1963.

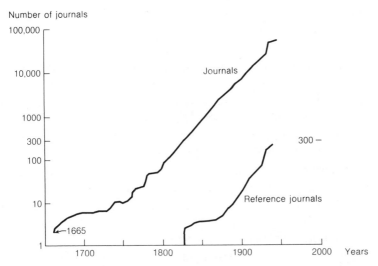

Figure 14.1. Growth in the total number of scientific journals.

entific research and experimental design work could be dispensed with if information on similar work already done were available. The resulting losses in the United States have been $1.25 billion. According to G. N. Dobrov, in 1946 40 percent of applications for invention certification in the area of coal-combine construction were rejected as repetitious. In 1961 this figure had risen to 85 percent.

■ THE FORMALIZATION OF SCIENTIFIC LANGUAGE

CAN WE CONCLUDE from this that there is an information crisis in science? It is perhaps too early to speak of a crisis, but we can already see that as a result of the continuous growth in the stream of information there will be a crisis in the near future if qualitative changes do not take place in the organization of scientific research. Until now scientific research has been organized in forms which developed traditionally, by themselves. Not only are they not the result of scientific investigation, but until recently they have not even been a subject of investigation. So there must be a scientific approach to the problem of organizing scientific activity—that is, a new metasystem transition: scientific control of the system of science. This metasystem transition has two aspects. The first, which does not go beyond the framework of science as a subsystem in the system of culture, creates a new level of the hierarchy within the framework of science as a primarily linguistic activity. This is what we called metascience in the preceding chapter. The second aspect concerns science as a social phenomenon. This aspect has come to be called the science of science [in Russian, *naukovedenie*].

We introduced the concept of the metascience without having connected it to the information problem. When speaking of mathematics, however, we remarked that the metascientific, conceptual approach is the organizing principle for the limitless number of theories and problems axiomatic symbolic mathematics can generate. The connection with the information problem in the natural and technical sciences is obvious here. There is a great deal that can be investigated, and many research plans can be boldly outlined. But one must have first clear planning principles, plans for plans. Otherwise there will be anarchy among plans, and when anarchy occurs the decisive factors are frequently those remote from the interests of science: considerations of prestige, personal contacts, and the like. Furthermore, it is essential for the language of the natural sciences and engineering to be completely formalized; then the aggregate of human knowledge will appear in the form of a harmonious system; only then will it become possible to work out the scientific principles of planning science. One should not think that the process of formalization is something "formal," that is to say syntactical and amounting to nothing but new notations. The problem of formalization of the scientific language is a conceptual, semantic problem. It is the problem of working out new concepts, a problem which resembles the formalization and axiomatization which occurred in mathematics.

A completely formalized language is a language accessible to the machine. When the edifice of science has a formalized frame we can separate the work that can be done by machines and automata from the jobs that require creative human participation. After the separation the machine work can be assigned to machines. Today, of course, the very simplest tasks of this sort are already being done by machines (automation and the use of computers), but formalization will make it possible to raise considerably the level of problems solved by machines. This refers above all to the processing of information flows. Systematization and storage of information, selection of needed information, and very simple information conversions—these and other tasks which make up the information problem today cannot be satisfactorily resolved by machines without complete formalization of language. It is difficulties in formalizing language which at the present time limit the application of computers in information science. The advances that are being made in this area are primarily related to more or less successful formalization of more or less extensive parts of the scientific-technical language.

■ **THE HUMAN BEING AND THE MACHINE**

HOWEVER, turning over the lower levels of science to machines should involve, and already is involving, not only linguistic activity but also direct manipulation of the natural objects under study. Properly speaking, each time modern automation is used in scientific experiments it is indeed an "entry of the machine into research." Raising the level of automation in this or that particular sphere of research implies complete formalization of a corresponding part of the scientific language. Automatic scanning of photographs with traces (tracks) of elementary particles and sorting out given configurations of tracks is a prototype of future achievements in this area. The universal arrival of machines in direct contact with nature will require universal formalization of the language of science. The next stage which can be anticipated is independent machine formulation of experiments in accordance with metascientific recommendations.

As machines are increasingly used in science and production, the human being will become increasingly free from noncreative activity—which, no matter how paradoxical it may seem, becomes needed precisely because of the successes of creative activity! For what is creativity? Above all creativity is constructive action, action that leads to an increase in the level of organization in the world. But an action is not characterized as creative only on the basis of its results. These results must be considered within the relationship to the mechanism of the action or the relations between this action and the system that gave rise to it. The same action may be a creative act when it is done for the first time and mechanical repetition of the past when it is done according to established, known rules, by applying standard procedures. Nothing that is produced within the framework of an already existing system of control, whether it is work by a computer or the composition of stereotyped articles, is creativity. Creativity always goes beyond the framework of the system; it is free action. Creativity is a metasystem transition. The evolution of the universe is continuous creativity. One of the manifestations of this process is creative acts in culture which establish new levels of control and in this way deprive lower-level actions of their creative character. Thousands of slaves had to be driven to build a pyramid; thousands of arithmetic operations had to be performed to calculate the exact positions of the planets on paper. Machines will rid the human being of that sort of work and transfer human activity to that level of the hierarchy which is still creative at the given moment. With time, this level will also cease to be creative; the boundary between creative and uncreative work is steadily crawling upward.

Ideally, immediately after the discovery of the presence of a system in some activity, this activity (or the part subordinate to the system discovered) could be turned over to a machine. Unfortunately, there is at present a considerable gap between the time an uncreative component appears and the time when there is a practical possibility that it can be turned over to a machine. The development of automation in the realm of nonlinguistic activity, accompanied by formalization of language in the realm of linguistic activity, is lessening the gap, but it remains large. The information problem in science, the necessity of routine, stereotyped research, and the need to overcome organizational difficulties to conduct experiments are all evidence this gap exists in scientific activity. In production, we are still a long way from automatic plants capable of producing motor vehicles and television sets according to plans fed to them. We are even farther from the time when there will be nothing but automatic plants. But sooner or later this will occur. The gap will be eliminated or reduced to a minimum. The formalization of language and automation will rid human beings of uncreative work just as the use of mechanical energy has for the most part rid us of heavy physical labor.

■ SCIENTIFIC CONTROL OF SOCIETY

THE SOCIAL ASPECT of the problem of controlling science is inseparable from the problem of controlling society as a whole. Science and production are growing into a single system, and politics and ideology are also inseparably linked to it. Furthermore, both aspects of the metasystem transition necessary for the development of science (the metascientific and social aspects) are also inseparably linked, and there is no hope of fully carrying out the former without carrying out the latter. Thus we have here essentially a single problem—the problem of scientific control of society. And even from the point of view of "pure" science this problem is the principal one; progress is impossible unless it is solved.

In the initial stages of the development of science, scientists had a comparatively proper justification for nonintervention in the practical affairs of society. It was possible to say that science itself was one of the highest values of existence and would demonstrate its amazing capabilities in the future; in its embryonic state, it would have to be given the peace and warmth needed for development, no matter what. The scientist could say, like a hen sitting on her eggs: "Do what you want, but just leave me in peace! I am hatching a remarkable chick. That is the main thing."

In our day this sort of reasoning is pure hypocrisy. The remarkable chick has come out of its shell and requires food. To isolate it from the environment now would mean to starve it to death.

■ SCIENCE AND MORALITY

THUS SCIENCE CLAIMS the role of supreme judge and master of the entire society. But will it be able to handle this role? After all, people need not only knowledge of the laws of nature and the ability to use them. They also need certain moral principles, answers to such questions as what is good and what is bad? What should a person strive toward and what should a person oppose? What is the meaning and goal of the existence of each person and of all humanity?

Strictly speaking, science cannot answer these questions. The ideas of the good, the goal, and the duty which are part of moral principles are beyond the bounds of science. Science engages in the construction of models of that reality which actually exists, not that which should be. It answers the questions: What really is? What will be if such-and-such is done? What must be done so that such-and-such will be? But science cannot in principle answer the question "What must be done?" without any "if" or "in order that." As a certain American philosopher remarked, no matter how much you study the train schedules you will not be able to choose a train if you do not know where you are going. All attempts to construct moral principles on a scientific basis inevitably lead in the end to the question "What is the Supreme Good?" or "What is the Supreme Goal?" which are essentially the same thing. Scientific knowledge and logical deductions are relevant to moral problems only to the extent that they help deduce answers to particular questions from the answer to this general, final question. The problem of the Supreme Goal remains outside science and its solution necessarily requires an act of will; it is in the last analysis a result of free choice.

This in no sense means that science has no influence at all on the solution to this problem. True to its principle of investigating everything in the world, science can look from outside at the human being and at entire societies which are deciding the problem of the Supreme Goal for themselves. Science can analyze various aspects of this situation and predict the results to which adoption of a particular decision will lead. And this analysis can significantly influence the process of solving the problem, although it does not change the nature of the solution as a freely made choice.

■ THE PROBLEM OF THE SUPREME GOOD

WHEN AND HOW does the problem of the Supreme Good and the Supreme Goal emerge? It is obvious that the animals did not have it, nor was it found in the early stages of the development of human society. Until a certain time, good for both human beings and animals was that which brought satisfaction, and there was a hierarchy of goals—crowned by the instincts for preservation of life and continuation of the species—that corresponded to the hierarchy of satisfactions. The concept of the goal and the concept of the good are, in general, inseparable; they are two aspects of a single concept. The human being strives toward good, by definition, and calls that toward which he strives good. In the stage when good is equated with satisfaction the human being does not differ in any way from the animal in a moral sense; for the human being, moral problems do not exist. The point here is not the nature of the satisfaction, but the fact that it is given, that the criterion of satisfaction is the highest controlling system—one that changes goals but that does not undergo changes itself. Even from a purely biological point of view human satisfactions differ from animal satisfactions. As an example we may recall the sense of the beautiful. And as the social structure becomes more complex the human being acquires new satisfactions which are unknown to animals. Nonetheless, this does not create the problem of the Supreme Good. That arises when culture begins to have a decisive effect on the system of satisfactions, when it turns out that what people think, say, and do is capable of changing their attitude toward the world to such an extent that events which formerly caused satisfaction now cause dissatisfaction, and vice versa. True, satisfactions at the lowest level (those deriving from direct satisfaction of physical needs) hardly change at all as culture develops, but satisfactions of the highest level (elation at one's skill in hunting, physical endurance, and the like) are sometimes capable of outweighing low-level dissatisfaction. In this way the criterion of satisfaction itself proves subject to control. A metasystem transition occurs; the social scale of values and system of norms of behavior emerge.

But this is only the prologue to the problem of the Supreme Good. In primitive society the norms of behavior can be compared to animal instincts; in the social super-brain they are in fact a precise analogue of the instincts embedded in the brain of the individual animal. Control of association (thinking) destroys instincts or, to put it better, it demotes them and puts social norms of behavior in the topmost place. In primitive society these norms are just as absolute as instincts are for the animal. And although they do change in the process of society's development, just as instincts change in the process of evolution of the species, this is unconscious change. They are perceived by each individual as something given and beyond doubt. But then one more metasystem transition occurs, the transition to critical thinking, and then the problem of the Supreme Good emerges in full.

Now people not only influence their own criteria of satisfaction through their linguistic activity, but they are *conscious* of this influence. The simple "I want it that way!" loses its primary, given quality. When a person becomes aware that what he wants is not only a result of his upbringing but also depends on himself and may be changed by reflection and self-education, he cannot help asking himself what he *should* want. In his consciousness he finds an empty place that must be filled with something. "Is there an absolute Supreme Good toward which one should strive?" he asks himself, "How should one live? What is the meaning of life?"

But he cannot get unequivocal answers to these questions. A goal can only be deduced from a goal, and if a person is free in his desires, then he is also free in his desires for desires. The circle of doubts and questions closes and there is nothing more to rely upon. The system of behavior is suspended in the air. Naive primitive beliefs and traditional norms of behavior collapse. The age of religious and ethical teachings arrives.

There are many of these teachings and they differ in many ways, but at the same time it appears that they also have a great deal in common, at least if we speak of the teachings which have become widespread. Our job now is to determine whether the scientific worldview leads us to some type of ethical teaching, and if it does, which one. At the same time we shall discuss the question of the nature of the common denominator of the different ethical teachings.

■ SPIRITUAL VALUES

BEFORE DISCUSSING the problems of the Supreme Good and the meaning of life we must gain assurance that the problem is worth discussing. There are many people whose point of view may be called the theory of natural values. According to this theory the creation of ethical teachings is an idle occupation if not a harmful one. This theory asserts that human nature contains, along with needs and instincts of animal origin, a yearning for specifically human spiritual values such as knowledge, beauty, justice, and love of one's neighbor. Achieving these values brings the highest satisfaction. The task of a human being is to develop these yearnings in himself and in others and thus obtain the highest satisfaction from life. This is the one natural goal of the human being, the one natural purpose. Philosophical religious and ethical teachings which begin from a priori principles or principles taken from who knows where can only muffle and distort these natural, truly human yearnings and force people to act basely in the name of a Supreme Good which they have invented.

What can we say about this theory? It is convenient as a pretext for avoiding the solution of a difficult question. It also has the merit of shunning extreme positions. But, unfortunately, it is untrue. It is contrived to a much higher degree than the other teachings which openly admit their dogmatic nature. The assertion that striving toward the highest spiritual values is part of human nature in its literal, exact sense contradicts the facts. Children carried off by animals who grow up away from human society do not show an understanding of the highest values of modern civilized people; they generally do not become full-fledged people. Therefore, there is nothing in the actual structure of the developing brain that would unequivocally generate those specific higher aspirations of which the theory of natural values speaks.

"Oh no!" a supporter of this theory will say, becoming terribly indignant at such a vulgarization of his views. "We are certainly not speaking of the concrete ways these yearnings are manifested; what we refer to is their general foundation, which requires the conditions created by society if it is to manifest itself."

But then the theory of natural values commits the sin of switching concepts. To say "general foundation" is to say nothing if we do not give the concrete substance of this foundation and its connection with observed manifestations. From the point of view being developed in this book, the general foundation of the highest values recognized at the present time by a majority of the human race really does exist; it is inborn, encoded in the structure of the genes of each human being. This foundation is the ability to control the process of associating. It may be tentatively called the "knowledge instinct" (see chapter 4), but this is just a figurative expression. The profound difference between this ability and instinct is that instinct *dictates* forms of behavior while control of associating mainly *permits* them and removes old prohibitions. Control of associating is an extremely undifferentiated, multivalued capability which admits diverse applications. Even what we call thinking is not an inevitable result. And what can we say about the more concrete forms of mental activity?

Control of associating is more a destructive than a constructive principle; it needs constructive supplementation. This supplementation is the social integration of individuals, the formation of human society. It is in the process of development of society that spiritual values originate. Of course, they are far from accidental, but it is a long way from their general foundation implanted by nature in all human beings to spiritual values, and on this road it is the logic of society, not the logic of the individual, that governs. This road is not unambiguous and it is not complete.

The theory of natural values, in speaking dimly of the "general foundation" of spiritual values, thus actually equates certain particular ideals recognized at the present time by some (possibly many) people with this "general foundation" which is absolute, invariable, and implanted in human nature. Two consequences follow from this error. For one, the theory of natural values does a disservice to the spiritual values it promotes when it promotes them on a false basis. It is like the well-wisher who started defending the right of a peasant lad to human dignity not on the basis of the general principles of humanism but rather by attempting to prove his noble origin; the deception can easily be revealed and the unfortunate young man will be flogged. In the second place, this theory does not contain any stimuli to the development of spiritual values; it is antievolutionary, conservative to an extreme.

What do we have in mind when we say that some particular values are natural for the human being? Obviously we mean that they are dictated, established for human beings by nature itself. For the animal, instincts are the goals which nature gives him, and what fits the instincts is natural for him. But nature does not give the human being goals; the human being is the highest level of the hierarchy. This is a medical fact, as Ostap Bender [3] would say, a fact of the organization of the human brain. The human being has nowhere from which to receive goals; he creates them for himself and for the rest of nature. For the human being there is nothing absolute except the absence of absolutes and there is nothing natural except endless development. Everything that seems natural to us at a given moment is relative and temporary. And our current spiritual values are only mileposts on the road of human history.

It is worth thinking about the meaning of life. To think about the meaning of life means to create higher goals and this is the highest form of creativity accessible to the human being. This type of creativity is always needed because the highest goals must change in the process of development and will always change. And each person must somehow decide this question for himself since nature has given him such an opportunity. Assurances that this problem has been solved or assurances that it is insoluble are lies which some use deliberately; others fall back on them from mental laziness and lack of fortitude. The question is, of course, insoluble at the level of pure knowledge; it must include an element of free choice. But conscious choice accompanied by study of the object and reflection is one thing and blind imitation of an example imposed upon us is something else. In one way or another someone creates the highest goals, because outside of society, "in nature," there are none. Every person is given this capability to some extent; to voluntarily reject the use of it is the same thing as for a healthy animal to voluntarily reject physical movement and use of the muscles.

■ THE HUMAN BEING IN THE UNIVERSE

THE CRITICISM of the theory of natural values shows clearly that element of the scientific picture of the world we can use as a starting point to arrive at definite moral principles, or at least definite criteria

---

[3] Hero of the novel *Twelve Chairs* by Ilf and Petrov—trans.

for evaluating them. This element is the doctrine of the evolution of the universe and the human role in it. And so, let us set off.

The assertion of the continuous development and evolution of the universe is the most important general truth established by science. Everywhere we turn we observe irreversible changes subordinate to a majestic general plan or to the basic law of evolution, which manifests itself in the growing complexity of the organization of matter. Reason emerges on Earth as a part of this plan. And although we know that the sphere of human influence is a tiny speck in the cosmos still we consider the human being the crown of nature's creation. Experience in investigating the most diverse developing systems shows that a new characteristic appears first in a small space but, thanks to the potential enclosed in it, engulfs a maximum of living space over time and creates the springboard for a new, higher level of organization. Therefore we believe that a great future awaits the human race, surpassing everything that the boldest imagination can conceive.

But no one person is the human race. What can a person say about himself, about the place of his own mortal self in the universe? What can the human being attain? How do one's will and consciousness enter the scientific picture of the world?

One hundred years ago the portrait of the world that science depicted was completely deterministic. If one took it seriously, one could become an absolute fatalist. But we know now that this picture was wrong. According to contemporary notions the laws of nature are exclusively probabilistic. Events may be more or less probable (or completely impossible), but there is no law that can force events to flow in a strictly determined manner. The laws of nature more often demonstrate the impossibility of something than the reverse; it is not accidental that the most general laws are prohibitive (the law of conservation of energy, the law of increasing entropy, and the uncertainty relation). Cases where the course of events can be predicted quite accurately far into the future are more the exception than the rule—an example here is astronomical predictions. But they are possible only because we encounter here an enormous difference in time scales between astronomical and human time. If we were to approach the motions of the celestial bodies with the time scales inherent in them it would turn out that the only predictions we could make would be as limited as our predictions regarding the molecules of air we breathe. So the successes of celestial mechanics which inspired Laplace in his formulation of determinism are a very special case.

Indeterminacy is deeply implanted in the nature of things. The evolution of the universe is a continuous and universal elimination of this indeterminacy, a continuous and universal choice of one possibility from a certain set of possibilities. We can compare two situations involving choice—extreme cases that have been well-studied.

The first situation is the collision of two elementary particles. Knowing the initial conditions of the collision, we can give the probability of particular results, but nothing more. For example, if the probabilities that a colliding particle will be deflected upward and downward are identical, we cannot now—and never shall be able to—predict in which direction the particle will go. Nonetheless, nature makes its choice. This act of choice, which is among the most elementary, is according to modern notions a blind one. Changes in the evolution of the universe occur only because of the interweaving and play of an infinite number of such acts.

The second situation is the act of will of the human personality. We can study this act from outside, just as we study the collision of particles. This is the basis of behavioral psychology. If we know the conditions in which a person is placed and some of his psychological characteristics, we can make some predictions, also purely probabilistic. But when we view this situation from within—as our own free choice (as an act of manifesting our personality)—what had appeared unpredictable in principle when considered from outside is now seen as free will.

The nature of the unpredictability in these acts is the same, as is the impossibility of watching the system without affecting it; but how greatly they differ in their significance! The act of will encompasses an enormous space-time area as compared to the act of the scattering of particles. In addition, the act of will may be a creative act, not the blind, inert material of cosmic evolution but its direct expression, its moving force.

■ THE DIVERGENCE OF TRAJECTORIES

ALL THE SAME, the human being is extraordinarily small in comparison not only with the universe, but with the human race as a whole, and this again inclines us to think of the insignificance of the act of individual will and the law of large numbers would seem to reinforce us in this thought. We must note that superficially understood and incorrectly applied scientific truths very often promote the acceptance of false conceptions. That is how things are at present. Relying on the law of large numbers people reason as follows. There are 3 billion people on Earth. The destiny of the human race is the result of their combined actions. Because the contribution of each person to this sum is equal to one three-billionth no one person can hope to significantly affect the course of history, not even accidentally. Only general factors which influence the behavior of many people simultaneously count.

In reality this reasoning contains a flagrant error, because the law of large numbers is only applicable to an aggregate of independent subsystems. It could be applied to the human race if all 3 billion people acted with absolute independence and knew absolutely nothing about one another. However, as the human race is a large and strongly interconnected system, the acts of some people have very great effects on the acts of others. In general such systems possess the characteristic of divergence of trajectories, which is to say that small variations in the initial state of the system become increasingly larger over time. We call the situations in which the law of divergence of trajectories manifests itself in an unquestionable, obvious way crises. In a crisis situation enormous changes in the state of the system depend on minute (on a system scale) factors. In such a situation the actions of one person, possibly even a single word spoken by the person, may be decisive. We are inclined to consider crisis situations rare, but we know many constantly operating factors that multiply the influence of a single person many times over. These are the so-called *trigger mechanisms*. Only a very slight effort is required to press the trigger or control button, but the consequences resulting from this action may be enormous. It is hardly necessary to say how many such mechanisms there are in human society.

Nonetheless, the idea of the little person, this fig leaf with which we conceal in front of others the shame of our cowardice, does not give up without a struggle. Most people, the "little person" says, do not participate in crisis situations and do not have access to triggers.

Many people will perhaps recall the rhyme which ends with the words:

> For want of a battle the kingdom was lost—
> And all for want of a horseshoe nail.

The rhyme describes a trigger mechanism which goes from a slipshod blacksmith who did not have a nail to the defeat of an army. We take this story as humorous, not wishing to see it as completely serious. However, our entire lives consist of such multi-stepped dependencies. Mathematical investigation of large interconnected systems shows the same thing: trajectories diverge. An initially insignificant deviation (the lack of a nail in the blacksmith shop) enlarges step by step (the shoe falls off, the horse goes lame, the commander is killed, the cavalry are crushed, and the army flees). But we take a skeptical attitude toward such long chains because in our everyday life we are almost never able to trace them reliably from start to finish. In the first place, each connection between links of the chain is probabilistic: a lame horse certainly does not necessarily doom the commander. In the second place, following the relationship of events constantly raises questions of the type "What would have happened if . . .?" It is hard to find two people who give the same answers to a series of such questions, but it is impossible to turn the clock back and look. Finally, we practically never have the necessary information.

But that we cannot trace these chains in the opposite direction should not eclipse our awareness of their existence when we think about the consequences of our actions. Crisis situations are rare not because small factors rarely have major consequences (they do), but rather because we are seldom fully aware of the chain of events. We can never foresee the results of our actions exactly. The only thing available to us is to establish general principles through whose guidance we increase the probability of Good, that is, the probability of those consequences which we consider desirable. We should act in accordance with these principles, viewing each situation as a crisis situation because the importance of each act of our will may be enormous. By always acting in such a way we unquestionably make a positive contribution to the cause of Good. Here the law of large numbers operates at full strength.

■ ETHICS AND EVOLUTION

BUT WHAT IS GOOD? What are the Supreme Good and the Supreme Goal? As we have already said, the answer to these questions goes beyond the framework of pure knowledge and requires an act of will. But perhaps knowledge will lead us to some certain act of will, make it practically inevitable?

Let us think about the results of following different ethical teachings in the evolving universe. It is evident that these results depend mainly on how the goals advanced by the teaching correlate with the basic law of evolution. The basic law or plan of evolution, like all laws of nature, is probabilistic. It does not prescribe anything unequivocally, but it does prohibit some things. No one can act against the laws of nature. Thus, ethical teachings which contradict the plan of evolution, that is to say which pose goals that are incompatible or even simply alien to it, cannot lead their followers to a positive contribution to evolution, which means that they obstruct it and will be erased from the memory of the world. Such is the immanent characteristic of development: what corresponds to its plan is eternalized in the structures which follow in time while what contradicts the plan is overcome and perishes.

Thus, only those teachings which promote realization of the plan of evolution have a chance of success. If we consider the cultural values and principles of social life which are generally recognized at the present time from this point of view, we shall see that they are all very closely connected with our understanding of the plan of evolution and in fact can be deduced from it. This is the common denominator of the ethical teachings which have made a constructive contribution to human history.

But there is still a great distance between this objective and unbiased view of ethical principles and the decision to follow them. Really, why should I care about the plan of evolution? What does it have to do with me?

■ THE WILL TO IMMORTALITY

A VERY IMPORTANT FACT—that human beings are mortal—now must be considered. Awareness of it is the starting point in becoming human. The thought of the inevitability of death creates a torturous situation for a rational being and he seeks a way out. The protest against death, against the disintegration of one's own personality, is common to all people. In the last analysis, this is the source from which all ethical teachings draw the volitional component essential to them.

Traditional religious teachings begin from an unconditional belief in the immortality of the soul. In this case the protest against death is used as a force which causes a person to accept this teaching; after all, from the very beginning it promises immortality. If immortality of the soul is accepted then the stimulus to carry out the moral norms imposes itself: eternal bliss for good and eternal torment for bad. Under the powerful influence of science the notions of immortality of the soul and life beyond the grave, which were once very concrete and clear, are becoming increasingly abstract and pale, and old religious systems are slowly but surely losing their influence.

A person raised on the ideas of modern science cannot believe in the immortality of the soul in the traditional religious formulation no matter how much he may want to; a very simple linguistic analysis shows the complete meaninglessness of this concept.

The will to immortality combined with the picture of the world drawn above can lead him to just one goal: to make his own personal contribution to cosmic evolution, to eternalize his personality in all subsequent acts of the world drama. In order to be eternal this contribution must be constructive. Thus we come to the principle that the Highest Good is a constructive contribution to the evolution of the universe. The traditional cultural and social values may be largely deduced from this principle. To the extent that they conflict with it they should be cast aside as ruthlessly as we surpress animal instincts in the name of higher values.

The human being continues somehow to live in his creations:

No! All of me will not die! In the cherished lyre my soul
Will survive my ashes, it will not decay.
(PUSHKIN, *"I Have Raised a Monument to Myself,"* 1836)

What is the soul? In the scientific aspect of this concept it is a form or the organization of movement of matter. Is it so important whether this organization is embodied in the nerves and muscles, in rock, in letters, or in the way of life of one's descendants? When we try to dig down to the very core of our personality, don't we come to the conviction that its essence is not a repeating stream of sensations or the regular digestion of food, but certain unrepeatable, deeply individual creative acts? However, the physical result of these acts may go far beyond the space-time boundaries of our biological body. Thus we begin to feel a profound unity with the Cosmos and responsibility for its destiny. This feeling is probably the same in all people, but it is expressed differently in various religious and philosophical systems. It is this feeling that art teaches which elevates the human being to the level of a cosmic phenomenon.

Thus, the scientific worldview brings us to ethics, which points out the Supreme Values and demands that we be responsible for and actively pursue them. Like any ethics it includes the act of will, which we have called the will to immortality. If a person cannot or does not want to perform this act, then no knowledge, no logic will force him to accept the Supreme Values, to become responsible and active. And God save him! The Philistine who has firmly resolved to be content with his wretched ideal, who has resolved to live as a humble slave of circumstances, will not be elevated by anything and will pass from the stage without a trace. The person who does not want immortality will not get it. Just as the animal deprived of its instinct for reproduction will not perform its animal function, so the human being deprived of the will to immortality will not fulfill his or her human function. Fortunately, this case is the exception, not the rule. The will to immortality is not the privilege of certain "great" people, it is a mass characteristic of the human being, a norm of the human personality which serves as the source of moral strength and courage.

How convincing and acceptable will the ethical ideals we have deduced from the scientific worldview be for a broad range of people, our contemporaries and descendants? Doesn't all this reasoning sound a little too abstract and unfeeling? Is it capable of involving, of affecting the emotions? It is, and this is shown by many examples. The ideas of evolution and personal participation in the cosmic process conquer the imagination; they give life depth and meaning. But in return they demand bold conclusions and a readiness to sacrifice the conventional and adopt the unexpected and uncanny if that is where logic inexorably leads.

It is natural to expect that those who are engaged in science will have a positive attitude toward construction of an ethical system on the basis of the scientific worldview. This expectation is for the most part borne out. The scientists have many "fellow travelers" too. But there are also many enemies or, at least, persons who do not wish us well. In some circles (especially among the intelligentsia in the humanities) it is fashionable to curse scientists for their "scientism," their endeavors to construct all life on a scientific basis, surreptitiously substituting science for all other forms of spiritual life. These attitudes (which can hardly be called justified) are engendered primarily by fear in the face of that unknown future toward which the development of science is inexorably (and rapidly!) drawing us. The fear is intensified by misunderstanding, for neither the broad public nor the representatives of the intelligentsia in the humanities and arts ordinarily understands the essence of modern scientific thinking and the role of science in spiritual culture. This problem was set forth brilliantly by C. P. Snow in his 1956 lecture entitled "The Two Cultures." [4]

[4] C. P. Snow, *The Two Cultures and the Scientific Revolution* (London: Macmillan, 1959).

Science to the modern person is what fire was to the primitive. And just as fire aroused a whole range of feelings in our ancestors (terror, amazement, and gratitude), so science today arouses a similar range of feelings. Fire has an attractive and enchanting force. The primitive looked at fire and delights and dim premonitions earlier unknown rose in his soul. It is the same with science. Science fiction, for example, is just like the visions of primitives sitting around a fire. And constructing supreme goals and principles on the basis of the scientific picture of the world can be called fire worship. These metaphors do not degrade; they honor modern fire worshipers. After all, we are very deeply indebted to the imagination of our ancestors who were enchanted by the dancing flames of the fire.

■ INTEGRATION AND FREEDOM

THE PROCESS of social integration has never gone on so furiously and openly as it does today. Modern science and engineering have put every person in the sphere of influence of every other. Modern culture is global. Modern nations are enormous mechanisms which have a tendency to regulate the behavior of each citizen with increasing rigidity—to define needs, tastes, and opinions and to impose them on people from without. Modern people are hounded by the feeling that they are being turned into standardized parts of this mechanism, and are ceasing to exist as individuals.

The basic contradiction of social integration—that between the necessity of including the human being in the system, in the continuously consolidating whole, and the necessity of preserving the individual as a free, creative personality—can be seen today better than ever before. Can this contradiction be resolved? Is a society possible which will continue to move along the path of integration but at the same time ensure complete freedom for development of the personality? Different conceptions of society give different answers.

The optimistic answer to the question sounds positive. Each successive stage in the integration of society will probably involve some external limitations not fundamental from the point of view of creative activity. On the other hand, each stage will foster a liberation of the nucleus of the personality, which is the source of creativity. Belief in the possibility of such a society is equivalent to belief that the impulse implanted by nature in the human being has not been exhausted, that the human being is capable of continuing the stage of cosmic evolution he has begun. After all, the personal, creative principle is the essence of the human being, the fundamental engine of evolution in the age of intellect. If it is suppressed by social integration, movement will stop. On the other hand, social integration is also essential. Without it the further development of culture and increasing human power over nature are impossible; the essence of the new level of organization of matter lies in social integration. But why should we suppose that social integration and personal freedom are incompatible? After all, integration has been successfully carried out at other levels of organization! When cells join into a multicellular organism they continue to perform their biological functions—exchange of matter and reproduction by division. The new characteristic, the life of the organism, does not appear despite the biological functions of the individual cells but rather thanks to them. The creative act of free will is the "biological" function of the human individual. In the integrated society, therefore, it should be preserved as an inviolable foundation and new characteristics must appear only through it and thanks to it.

If we refuse to believe in the possibility of an organic combination of social integration and personal freedom then we must give one of them preference over the other. The preference for personal freedom leads to the individualistic conception of society, while preference for social integration leads to totalitarian regimes.

Individualism views society as nothing more than a method of "peaceful coexistence" of individuals and increasing the personal benefits for each of them. But by itself this idea is inadequate to build a healthy society. Pure individualism deprives the life of a person of any higher meaning and leads to cynicism and spiritual impoverishment. In fact, individualism exists only thanks to an alliance with traditional religious systems—or, to put it better, by living as a parasite on them—because they are in principle hostile to individualism and permit it only as a weakness. With the collapse of the religious systems this parasite reaches enormous size. Individualism becomes a fearsome ulcer eating up society and inevitably, as a protest against itself, it gives rise to its negation, totalitarianism.

For totalitarianism, integration is everything and the individual is nothing. Totalitarianism constructs a hierarchical state system which is usually headed by one person or a small group of people. An ideological system is also constructed which each citizen is obliged to accept as his or her personal worldview. Anyone refusing to do this is subject to punishment, which may go as far as physical extermination. The person trapped in between the two systems becomes a thoughtless, soulless part in the social machine. The person is given only what freedom is necessary to carry out instructions from above. Every manifestation of individual activity is viewed as potentially dangerous to the state. Personal rights are abolished.

Striving to preserve and strengthen itself, the totalitarian state uses all means of physical and moral influence on people to make them suitable to the state—"totalitarian" people. The fundamental characteristic of the totalitarian person is the presence of certain prohibitions he is unable to violate. He may be a scientist, an investigator filled with curiosity, but upon approaching certain aspects of life his curiosity suddenly begins to evaporate. He may be a brave man, capable of giving his life for his country without a thought, but he trembles in fear before his leader. He may consider himself an honest man but speak what he knows to be a lie, and not connect this lie with his supposed honesty. He may steal, commit treason, and kill in the confidence that "it is necessary"; he will never permit himself to ask if it really is necessary. And he will walk a mile to avoid anything that might force him to think about this.

The totalitarian person is compensated for these tabus, which are imposed on precisely what constitutes the highest value of human existence, by the feeling of unity—the feeling that he belongs to an enormous aggregate of people who are organized into a single whole. The human being has an inherent, internal need for social integration, and totalitarianism's strength is that it plays on this need and satisfies it to some extent. The strength and danger of totalitarianism are that it stands for social integration, and social integration is an objective necessity.

But the totalitarian state is not the solution to the problem of social integration. It achieves wholeness by smoothing out differences among its constituent human units to the point where they lose their human essence. It cuts off people's heads and forces the stumps to be elated at the unity achieved at such a price. Totalitarianism is a tragically clumsy and unsuccessful pseudosolution; it is the abortion of social integration. By destroying the individual person it deprives itself of the source of creativity. It is doomed to rot and decay.

While individualism generates totalitarianism, totalitarianism, inversely, generates individualism. "Down with the collective!" cries the person raised in totalitarianism who has become aware of his slavery. "Leave me alone! I don't want unity! I don't want military might! I don't want a feeling of comradeship! I want to live the way I like! I! I! I!" Fearing punishment, however, he only imagines he is shouting this; at most he whispers it. His ego, which has grown up under totalitarian conditions, is a wretched, half-strangled one. And he becomes a purposeless Philistine with the perspective of a chicken. He is not interested in anything except his own self. He does not believe in anything and therefore he subordinates himself to everything. This is no longer a totalitarian personality, it is a miserable and cowardly individualist living in a totalitarian state.

Individualism and totalitarianism are two opposites linked in a common chain. There is only one way to break this circle: to set as our task conscious social integration with preservation and development of creative personal freedom.

■ QUESTIONS, QUESTIONS . . .

ATTEMPTS TO LOOK even farther, as far as imagination permits, produce more questions than answers.

How far will integration of individuals go? There is no doubt that in the future (and perhaps not too far in the future) direct exchange of information among the nervous systems of individual people (leading to their physical integration) will become possible. Obviously the integration of nervous systems must be accompanied by the creation of some higher system of control over the unified nerve network. How will it be perceived subjectively? Will the modern individual consciousness, for which the supreme system of control will be something outside and above the personal, something alien and not directly accessible, be preserved unchanged? Or will physical integration give rise to qualitatively new, higher forms of consciousness that will form a process that can be described as merging the souls of individual people into a single Supreme Soul? The second prospect is both more probable and more attractive. It also resolves the problem of the contradiction between reason and death. It is difficult to tolerate the thought that the human race will always remain an aggregate of individual, short-lived beings who die before they are able to see the realization of their plans. The integration of individuals will make a new synthetic consciousness which is, in principle, immortal just as the human race is, in principle, immortal.

But will our descendants want physical integration? What will they want in general? And what will they want to want? Already today the manipulation of human desires has become a phenomenon that cannot be discounted, and what will come in the future when the structure and functioning of the brain have been investigated in detail? Will the human race fall into the trap of the absolutely stable and, subjectively, absolutely happy society which has been described in the works of science fiction writers such as Zamyatin and Huxley?

To avoid falling into such a trap there must be guarantees that no control structure is the highest one finally and irreversibly. In other words, there must be guarantees that metasystem transitions will always be possible in relation to any system no matter how large it may be. Are such guarantees possible? Does consciousness of the necessity of the metasystem transition for development give people such guarantees? And is the very need for development, the yearning to continue development, ineradicable? We have reason to hope that it is. Having conquered the human consciousness, the idea of evolution seemingly does not want to go away. If we imagine that the human race will exist forever like a gigantic clock, unchanging and identical, with people (its machinery) being replaced as a result of the natural processes of birth and death, we become nauseous; this seems equivalent to the immediate annihilation of the human race. But will it always seem that way to our descendants? Perhaps now, when we feel that necessity of development, we should try to perpetuate this feeling? Perhaps this is our duty to the living matter which gave us birth? Suppose we have made such a decision. How can it be carried out?

Now let us pose the question of the pitfalls along the path of development in more general form. Ant society is absolutely stable. But that is not because it is poorly organized; the individuals which make it up are such that unifying them does not give rise to a new characteristic—it does not bring brains into contact (the poor things have virtually nothing with which to make contact). Is it possible for the remote descendants of the ants or other arthropods to become rational beings? Most likely it is not. It appears that the arthropods have entered an evolutionary blind alley, but perhaps we are in one too. Perhaps the human being is unsuitable material for integration and no new forms of organization and consciousness based on it will develop. Perhaps life on Earth has followed a false course from the very beginning and the animation and spiritualization of the Cosmos are destined to be realized by some other forms of life.

Let us assume that this is not true, that nature has not committed a fatal injustice in relation to the Earth. Now, when conscious beings have appeared, what should they do to avoid wandering unknowingly into a blind alley? For such a general question a general answer may be offered: preserve, even in some miniature, compressed form, the maximum number of variations; do not irreversibly cut off any possibilities. If evolution is wandering in a labyrinth, then when we come to a point where the corridors intersect and we choose the path going to the right we must not forgot that there is also a corridor going to the left and that it will be possible to return to this place. We must mark our path with ineradicable, phosphorescent dye. This is precisely the function of the science of history. But are the linguistic traces which it leaves adequate? Perhaps a conscious parallelism is essential in solving all social problems.

We shall hope that we have not yet made an uncorrectable mistake and that people will be able to create new, fantastic (from our present point of view) forms of organization of matter, and forms of consciousness. And then the last, but also the most disturbing, question arises: can't there exist a connection between the present individual consciousness of each human personality and this future superconsciousness, a bridge built across time? In other words, isn't a resurrection of the individual personality in some form possible all the same?

Unfortunately, all we know at the present time compels us to answer in the negative. We do not see any possibility of this. Neither is there a necessity for it in the process of cosmic evolution. Like the apes from which they originated, people are not worth resurrection. All that remains after us is what we have created during the time allotted to us.

But no one can force a person to give up hope. In this case there is some reason to hope, because our last question concerns things about which we know very little. We understand some things about the chemical and physical processes related to life and we also can make our way in questions related to feelings, representations, and knowledge of reality. But the consciousness and the will are a riddle to us. We do not know the connection here between two aspects: the subjective, inner aspect and the objective, external aspect with which science deals. We do not even know how to ask the questions whose answers must be sought. Everything here is unclear and mysterious; great surprises are possible.

We have constructed a beautiful and majestic edifice of science. Its fine-laced linguistic constructions soar high into the sky. But direct your gaze to the space between the pillars, arches, and floors, beyond them, off into the void. Look more carefully, and there in the distance, in the black depth, you will see someone's green eyes staring. It is the *Secret,* looking at you.

Index

Abacus, 195
Algebra, 247 ff.
Algorithmization, 248, 264, 267
Al-Kwarizmi, 200, 248-49
Analysis, logical, 137
Animism, 175
Appolonius, 242, 244-45, 253, 257
Archimedes, 242-44, 257
Archytas of Tarentum, 217
Aristotle, 217, 240
Ascheulian culture, 107
Associating, 60, 63, 67; control of, 328-29
Aurignac-Solutrean culture, 108
Axioms, 227 ff., 235, 277

Bacon, F., 289-90, 292
Beautiful (the), 86
Belief, primitive, 185
Bhascara, 251
Bohr, Niels, 301, 311
Bolyai, J., 278
Boole, G., 280
Boredom, 85
Bourbaki, N., 132, 273, 283, 287-88
Bowdich, T., 185
Branching of the penultimate level, law, 59, 89, 148, 216, 247
Brouwer, L. E. J., 284
Burali-Forti, C., 283

Cantor, G., 283-86
Cardano, G., 250-51
Catlin, G., 180

Cauchy, A., 275
Change, qualitative and quantitative, 105
Chellean culture, 107
Chukovsky, K., 77
Civilization, 174
Classifier, 25, 35, 183
Clavius, 275
Coding and decoding, 89
Codrington, R., 186
Communication channels, 12
Concept, 114, 238; particular, 22; Aristotelian, 22 fn., 117; abstract, 23 fn., 24, 171-72; general, 23 fn.; logical, 151
Conceptology, 313
Connectives, logical, 125, 149
Construct, 169, 256, 262, 269, 294
Copernicus, N., 296-97
Counting, 191-93
Creativity, 96,'321-22, 329, 339
Culture, material and spiritual, 101, 173
Cybernetics, xv, xvi-xvii, 2, 4, 55, 99

Deduction and induction, 292-95
Del Ferro, S., 249-50
Descartes, R., 244, 246, 251, 253, 257-62, 270-71, 273, 288, 291-92
Description: static and dynamic, 5; phenomonological, 53
Determinism, 304 ff
Diagrams: TOTE, 48, 65; structural and functional, 49-52; control, 50
Dialectic, 121
Diophantus, 248, 251-52, 257

Discriminator, 25
Dobrizhoffer, M., 184
Dobrov, G. N., 317 *fn.*, 319

Einstein, A., 167, 292-94
Ellis, A. B., 179, 182
Engels, F., 271
Equation, 208
Esthetics, social value of, 87-88
Euclid, 217, 225, 241, 252, 256, 277-78
Eudoxus of Cnidus, 217, 256
Euler, L., 262
Evolution, 98, 188, 236, 315, 322; basic law of, 1, 2, 330; precybernetic ("chemical") 2; first stage, 2; second stage, 3; third stage, 3; cybernetic, 4; of animal nervous system, 17, 30; centralization, 20; diagrams, 58, 96; era of reason, 74, 90; human emerges, 81; of brain, 108; moral significance, 330, 334-35; future, 343-44
Experiment: definition, 290; machine formulation of, 321

Feedback, 12
Fermat, P. de, 253, 257-59
Ferrari, L., 250
Fiore, 249
Formula, 210
Frank, P., 296-97
Free choice (free will), 141, 306, 329, 331
Frege, G., 280
Funny (the), 86

Galanter, E., 48, 54, 69
Galileo, G., 289
Gauss, K. F., 278
Goal, 42, 46

Hegel, G., 102, 121
Heisenberg, W., 301, 311
Hierarchy, 36, 43, 315: of nervous system, 21; definition, 26; of concepts, 28-30; origin in nature, 32; of visual concepts, 39; of classifiers, 44, 155; of goals, 47-48; in language, 152-53, 159, 169; of theories, 276-77
Hilbert, D., 282, 284-86

Hippasas, 216
Hippocrates of Chios, 216, 245
Huxley, A., 342
Huygens, C., 275
Hydra, 17-19

Identifier, 130, 151
Image, stabilized on eye, 38-40
Imagination, 79 ff
Incommensurable line segments, 240, 251, 254
Individualism, 339-41
Information, 10-13
Instinct, 69, 82, 325
Integration: social, 95, 338; physical, 342
Intuitionism, 284

Jones, G. H., 183
Journals, scientific, 318-19

Kant, I., 236
Knowledge, 73, 175, 185

Lambert, G., 278
Language: definition, 88; animal, 89; human, 90; two functions of, 94; material system of, 141; algebraic, 220, 251; mathematical, 223; geometric, 223; of symbols, 242; of modern mathematics, 253; formalized, 264 ff.; "language machine," 266-67; concrete and abstract, 268; formalization of scientific, 319-21
Laplace, P., 305-6
Learning (cognition), 68, 73, 83, 236
Leibnitz, G., 262, 273-74, 278, 280
Leroy, C., 99
Lettvin, J., 34 *fn.*
Level, of hierarchy, 26
Lévy-Bruhl, L., 176, 178, 186
Lobachevsky, N. I., 278, 279
Logic: mathematical, 123; its role, 143; formalized, 274

Magdalenian culture, 108
Mechanism (phil.), 305
Memory, 47
Metamathematics, 280, 282, 312
Metascience, 312, 320

Metasystem transition, ix-x, xv-xvi, 56 ff., 99,
    103-5, 111, 113, 167, 174, 177, 189, 216,
    247, 274, 282, 290-91, 312, 322, 342
Miller, G., 48, 54, 69
Model, linguistic, 92, 207
Models, modeling, 71, 92, 94
Mooney, J., 179
Mousterian culture, 107

Name and meaning, 88, 90, 92, 178 ff., 237,
    286
Neigebauer, O., 196
Nerve net, 14, 15-16
Neuron (nerve cell), 13-15; associative, 31
Newton, I. 167, 262, 293-94, 296-97, 305
Noise, 12
Noosphere, 99
Number notation: Greek, 193, 199; Egyptian,
    193, 201; Roman, 193; Kharoshti and
    Brahmi, 195; Babylonian, 196-99; Indian,
    200
Numbers: general, 170, 191 ff.; irrational,
    246, 251, 275; "unreal," 250; negative,
    250-51, 275; "imaginary," 251, 275;
    definitions, 255

Object, physical, 89, 130
Objectivization of time, 165
Organization in time, 5, 47

Pappas of Alexandria, 244, 253, 257
Paradox, Russell's, 283-85
Pavlov, I. P., 41, 60, 114
Peano, G., 280
Philosophy, 270-71
Pierce, C. S., 280
Plan of actions, 48
Planck constant, 300 ff., 310
Plato, 217, 224-25, 250, 256, 304
Platonism, Platonic ideas, 218, 237, 255-57
Play, 76
Pleasure of novelty, 85
Plutarch, 243
Predicate: general, 127, 151; primary, 139, 159
Pribram, K., 48, 54, 69
Price, D., 318 fn.
Pritchard, R., 38 fn.

Procul, 202
Proof, 212, 281
Ptolemy, 296-97
Pythagoras, 215, 216, 239

Quantifiers, 128
Quantity, geometric, 240
Quantum mechanics, 297, 299, 304, 306

Receptors and effectors, 15-16, 43
Recognition (discrimination), 22
Reflex: simple, 17; complex, 19-21, 55, 59;
    conditioned, 60, 67; unconditioned, 60
Regulation: definition, 42; origin, 43-45
Reichenbach, H., 139
Relativity: linguistic, 167; in physics, 167
Reliability, mathematical, 226
Replication, 2, 43-44
Representation, ix, 46, 60, 140, 183, 235, 297
Russell, B., 280, 283

Sapir, E., 160
Schnoll, S. E., 2 fn.
Schroeder, E., 280
Scientific method, 290 ff.
Self-knowledge, 92-93
Semantics, 135, 147
Semiotics, 135
Set theory, 283-86
Situation, definition of, 16
Snow, C. P., 337
Solomon, V., 184
Specialization, 258
Spirits, 181
Stairway effect, 102, 276, 291
State: of a system, 5; generalized, 50, 52, 62
Stevin, S., 200
Substance, 163-64
Super-being, 97, 105
Supreme Good, 324-28, 336
Syntax, 135
Systems: discrete and continuous, 6-10, 13;
    fully deterministic and reversible, 12

Tartaglia, N., 249-50
Taylor, E., 175
Teilhard de Chardin, 99-100